Leandro Buzzo Mourão Guimarães

Production of place by children with Autism Spectrum Disorder

Leandro Buzzo Mourão Guimarães

Production of place by children with Autism Spectrum Disorder

An analysis of municipal schools in Presidente Prudente -SP

ScienciaScripts

Imprint
Any brand names and product names mentioned in this book are subject to trademark, brand or patent protection and are trademarks or registered trademarks of their respective holders. The use of brand names, product names, common names, trade names, product descriptions etc. even without a particular marking in this work is in no way to be construed to mean that such names may be regarded as unrestricted in respect of trademark and brand protection legislation and could thus be used by anyone.

Cover image: www.ingimage.com

This book is a translation from the original published under ISBN 978-613-9-65817-6.

Publisher:
Sciencia Scripts
is a trademark of
Dodo Books Indian Ocean Ltd. and OmniScriptum S.R.L publishing group

120 High Road, East Finchley, London, N2 9ED, United Kingdom
Str. Armeneasca 28/1, office 1, Chisinau MD-2012, Republic of Moldova, Europe
Printed at: see last page
ISBN: 978-620-7-80044-5

Summary - atz numbers

GUIMARAES, Leandro Buzzo Mourâo

The production of the place of children with autism spectrum disorder (ASD) in municipal schools in Presidente Prudente -SP. Leandro Buzzo Mourâo Guimarâes - Presidente Prudente [s.n], 2017, 93p.

Supervisor: Raul Borges Guimarâes

Monograph (course completion work) - Paulista State University, Faculty of Science and Technology

Includes bibliography

1. construction of the feeling of place and non-place throughout the history of people with mental disorders; 2. autistic spectrum disorder: histories, concepts and definitions;

3. Qualitative methodologies; 4. Spatio-temporal relations and the production of a sense of place experienced by people with mental disorders; 5. A look at school inclusion for children with autism and its multi-scale meaning; 6. In conclusion.

Dedication

I dedicate this monograph to everyone who actively participated in its construction and to my parents for believing in my dream of becoming a geographer.

Thanks

I would first like to thank my parents, Angela and Sebastiâo, to whom I also dedicate this monograph, for making an effort every day to keep me in another city, in conditions that were often difficult.

I would also like to thank my advisor Raul Borges Guimarâes, who accepted me into the Biogeography and Geography of Health Laboratory (BioGeoS) from the first month of classes, believing in my object of study, always leaving his comfort zone, always trying to learn more every day. He showed me the way to the immensity of the sea, but he never let me drown, he bought my fights, he always fought by my side to give me the best he had. He also taught me not to be selfish, to take part in the lab's activities, and look, there haven't been many of them in the four years we've been together!!!!

I must also thank the staff at CETAS and the BioGeoS laboratory, who never refused to help anyone who didn't even know what research was. To NatiGeo, for helping me write a research project in my first year; to Patricia, for her meticulous Japanese mathematical methodology; to Carol and Mateus, for their help with research, studying and for their totally different views, which made us fight during these four years in the lab; to NatiBio, for her gentle, affectionate and suspicious way.

But I would especially like to thank some of the great friends I met in the lab for their immense partnership, in which we formed a "tough trio" that I will never forget in my life: Baltazar Casagrande, for his wonderful fieldwork, for giving me the opportunity to smell the bush and disconnect from my hectic life, for the gossip every afternoon in the lab, for his ogre-like manner, for his effort and field wisdom, I believe he is the geographer who knows the most about fieldwork! And to the other ogre, Matheus Moreno, for always being by my side, every day, with his crazy tics, making me feel embarrassed at balance parties, for the trips and the confidences.

When it comes to friends, I can't forget to mention my friend Valquiria, for the wonderful moments, the daily struggles we faced together, for never ceasing to believe in me, coming to my aid in difficult times and celebrating my victories together. To my classmates Gru, Diana, Ba Cardoso, Leticia, Alan, Maria, Sinthia, Gi.

And not least to Agda, who I had the opportunity to live with and for the moments of healthy food for lunch and beer for dinner, for the walks to Tigrào to let off steam, for the early morning tea, for the lemon popsicles, for the help with things at university, with the selection process for the master's degree and who was responsible for the wonderful cover.

Finally, all the teachers, especially Tagiane and Andréia, for their support with psychoanalysis; the staff at Unesp, and all the other colleagues who helped me immensely on this journey to become a geographer. To the relatives of the children with autism for the interviews, telling their stories and often making me emotional; to the school for the great opportunity to be with these children for two years.

At the boundary between the material world, where human activity takes place, and the imaginary world, opening up its symbolic content to the freedom of the spirit, we rediscover an inner, primitive geography, in which the original spatiality and profound mobility of man designate directions, trace the paths to another world (DARDEL, 2011, p. 5).

Summary

Autism Spectrum Disorder (ASD) is a pathology that presents disturbances in neurological development, with the main abnormal characteristics being impaired social communication and fixed or repetitive behaviors, usually affecting children by the age of two. In view of these characteristics, this project aims to contribute, from the point of view of health geography, to the discussion about school inclusion and the production of the place of children with ASD in the municipal schools of Presidente Prudente - SP. From the point of view of scientific initiation, it is essential to train geography students interested in the subject, which involves an interdisciplinary understanding.

Key words: Place, Autism Spectrum Disorder, School inclusion, Psychoanalysis.

Introduction

This monograph is organized into three main parts, written as simply as possible so that any reader can understand the subject. Each chapter also has a dialog box with key concepts, in order to give an introduction or more information on what the chapter is about.

The first part - Approaching the reader to the subject of the research - was divided into two parts, the first being the history of mental disorders and the second the history of Autism Spectrum Disorders, together with some geographical concepts.

The second part - Theory and methodologies - describes the main concepts underpinning the research. The methodological part was based mainly on qualitative research with interviews using a semi-structured script, participant observation, the doodle technique and the use of quantitative methodologies to support the analysis.

Finally, there are the Results and Conclusion, in Part III, which presents a scalar analysis, from the broadest aspects of the situation in Brazil, to the micro-analysis of the school and the subjects, increasing the detail as the scale decreases.

Part 1

Chap 1.

Key concepts:

***Place*: A** feeling of belonging to a place, or more, with or without a physical boundary. This feeling can be understood through actions that are projected in the mind, leaving our roots and security in these places (TUAN, 1983).

Ex[1] : My home is my place
Ex[2] : Brazil is my place

No Place: This concept adopted in the monograph means an aversion to the feeling of belonging, which is also felt materially. It differs from the concept of space expressed by Tuan (1983), because this space is already experienced with negative feelings.

Ex[3] : I have an aversion to the United States, so it's a no-go

Mental disorders: Mental disorders are often associated with significant suffering or disability that affects social, professional or other important activities. An expected or culturally approved response to a common stressor or loss, such as the death of a loved one, does not constitute a mental disorder. Social deviations in behavior (for example, of a political, religious or sexual nature) and conflicts that are basically about the individual and society are not mental disorders unless the deviation or conflict is the result of a dysfunction in the individual as described" (DSM-5, p. 20).

Place has always been the very essence of geographical science, according to the Humanist perspective. *"To reflect on place is to reflect on meaning in geography"*. In this view, place is thought of as a relationship of belonging, it is the place experienced as a warmth that we carry within us (OLIVEIRA, 2012, p.23).

Let's take the mirror as an example. We human beings are made up of appearance and essence, but when we look at our reflection and see ourselves in the mirror, our appearance is reflected, but our essence is not. Thus, we establish ourselves as places, with our essence, and when this is not transmitted to places, we start to build non-places, which are inhospitable, without identities, nothing more than appearances.

Places, therefore, can only be produced by those who live in them in the joint form of (physical) constructions, activities and meanings. Place extends to inextricable links with being (RELPH, 2012). Thus, place transcends materiality, but is not dissociated from it, since spatial marks and forms are not only based on the physical elements of objects

(MELO 2012).

When we look at subjects with mental disorders and places, we can see that there has been a clash throughout history, in which these subjects were considered incapable and useless to carry out any activities, being thrown out into the world, or forcibly isolated so as not to be seen by the norms of society. To the non-places. This conflict only began to be analyzed, especially by the humanities and social sciences, in the 1990s, with the broadening of health concepts and growing perspectives on empowerment and living with differences.

This new conception has stimulated public policies, enabling the construction of new places and relationships of belonging and research that expresses the theme, such as the ongoing project at the Faculty of Science and Technology (UNESP-fct) with the theme: "School Inclusion and Health Promotion: Producing the Social Place of People with Autism Spectrum Disorder in the municipal schools of Presidente Prudente - SP."

In this sense, in a more in-depth historical analysis, we can see that in previous periods in the western part of the planet, madness and the mad had other meanings, due to the great religious dominance of Christianity, being considered as demons, deified, a mistake of nature, among many other negative aspects. Their non-places were as diverse as possible: streets and ghettos, asylums and prisons, churches and hospitals (AMARANTE, 2007).

Pinel, in his "Medical-Philosophical Treatment of Mental Alienation or Mania" published in 1801, introduced several changes to the theoretical framework. For him, an important step had to be taken in the treatment of these people: *isolation from the outside world, in* which the alienated person could be treated in a way that suited their needs, thus removing the individual from external interferences that could harm their way of life and any difficulty in diagnosis by doctors regarding the illness, thus showing the *"PURE STATE" of* the disease (AMARANTE, 2007). Over the course of time, we see a change in the non-places based on Pinel's vision, with barriers being built around the non-places as a form of isolation from the outside world, resulting in little improvement.

Pessotti (1996, p.128) classifies this period, highlighted in the title of his work, as the "century of asylums", based on the rapid proliferation of this method throughout Europe in the 17th century *"[...] This is how the doctor becomes not only the organizer of the patient's (psychic) life, but also the agent of the social order, of the dominant morality"*

The anti-asylum struggle was long and resulted in very slow changes that have only gained more prominence in recent decades (AMARANTE, 2007). Several theorists have been important to this paradigm shift, in particular the work of FOUCAULT (1979). According to this French sociologist, one fact has long since become commonplace in sociology and mental pathology: illness only has the reality and value of illness within a culture that recognizes it as such (FOUCAULT, 1975). In Western culture, since its inception, the psychiatric hospital has been an institutional field for disputing standards of normality, with the empowerment of doctors, who became the authorities on madness, This was constituted through the techniques and procedures used for private or public interrogations, punishments with showers (cold and hot), moral preaching, repression, strict discipline, compulsory work, relationships of domesticity and servitude of the mentally ill (FOUCAULT, 1979).

Amarante (2007) adds that these measures were intended to re-educate the mind, dispel delusions and illusions and restore consciousness. For Foucault,

The medicine's mode of action follows a natural and discursive development, but the principle of its action is a proximity to the essence, an original communication with nature, an opening to a Principle. It is in this ambiguity that one must understand the successive privileges attributed during the 18th century to "natural" medicines, i.e. those whose principle is hidden in nature, but whose results are visible to a philosophy of nature: air, water, ether and electricity (FOUCAULT, 1972, p.336).

Foucault (1972) organizes some therapeutic ideas around curing madness. The first is established in the consolidation of madness, which would be the realization that subjects were unable to follow the natural course of their lives. Secondly, the purification and immersion that would result in the elimination of the impregnation of everything that would not be a pure state of nature. And finally, the regulation of movement, the fixation of ideas that prevail over others.

According to Pessotti's (1996) historical and bibliographical survey, these characteristics were brought with the arrival of the Royal Family in Brazil. This was a milestone in the history of madness in Brazil, especially with the founding of the first psychiatric hospital, the Hospicio D. Pedro II.

With the Great World Wars, there was a new social upheaval in which humanity began to take a different view of what had been happening to the mentally handicapped until then, both in terms of cruelty and solidarity, by analyzing that the conditions experienced in mental institutions were not very different from those offered in concentration camps, resulting in a lack of human dignity. However, despite the effort, the transformations were limited and remain unclear to this day (AMARANTE, 2007).

This new vision established in Europe, as in other periods, arrived late again. We can see that this movement in São Paulo was established around the 1970s, in a movement anchored in the Health Reform for Psychiatric Reform, which included the contribution of various players in society at the time.

Thus, the search for the transformation of the Mental Health system began to materialize, through the deinstitutionalization of a new way of dealing with madness and psychic suffering, becoming in 1990 a basic concept determining the reorganization of the service system, of Mental Health actions and a guideline of Mental Health policy. This led to the structuring of psychosocial care networks, in order to replace the model centered on hospitalization. Examples and components of these psychosocial restructuring networks are the CAPS (Psychosocial Care Center) and the RTs (therapeutic residences).

The real aim of the CAPS is to provide care for the population in its catchment area, providing clinical follow-up and enabling its users to reintegrate into society through access to work, leisure, exercising their civil and political rights, as well as strengthening their family and community ties, which demonstrates the importance of the role of the city as an instrument for fostering reintegration (SES, 2008).

According to the Secretary of Health of the State of São Paulo (SES, 2008), 45 psychiatric hospitals were closed in the State of São Paulo between 1989 and 2007, due to the circumstances of management and public policies, which began the process of deinstitutionalization of long-term resident patients, however, the State still has one of the highest concentrations of beds in the country. Until 2008, the state of São Paulo had 58 psychiatric hospitals, located in 38 municipalities in 15 Regional Health Departments (DRS), as shown by the Psychosocial Census (2008).

Figure 1 - Location of Psychiatric Hospitals in the State of Sao Paulo, 2008.

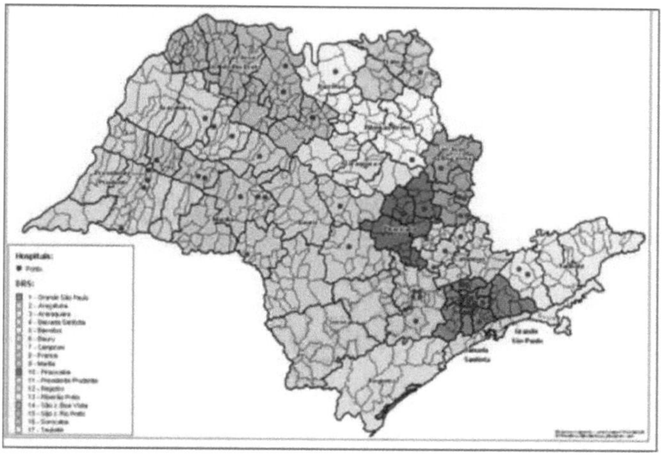

Source: SES - SP, 2008.

Of the 58 hospitals, 30 are managed by the state government and 28 are under municipal management. Of the state-run hospitals, 9 were public, 12 philanthropic and 9 private, while those under municipal management accounted for 22 philanthropic and 6 private.

In the 11th Regional Health Department, which has as its reference the municipality and city of Presidente Prudente - SP, considered the regional capital of Pontal do Paranapanema, 4 psychiatric hospitals were counted. They are:

1. Santa Maria Hospital in Pirapozinho. Located in the municipality of Pirapozinho, São Paulo, until 2008 it had 49 residents and 59 beds offered by SUS, 83% of which were occupied.

2. Sâo Joâo Hospital in Presidente Prudente. Until 2008, it had 88 residents and 180 beds offered by SUS, of which 49% were occupied.

3. Bezerra de Menezes Spiritist Psychiatric Hospital. By 2008, it had 116 residents and 225 beds offered by the SUS, 52% of which were occupied.

4. Clinica Psico Fisico Social Allan Kardec. Until 2008, it had 60 beds offered by SUS, but with the process of deinstitutionalizing the hospital, it was closed. The same process is currently taking place at the Bezerra de Menezes hospital, all three of which are located in the municipality of Presidente Prudente.

Table 1. Distribution of the Number of Residents in DRS XI, by Municipality and Hospital, 2008

Municipality	Hospital Name	Management	Nature	No. of residents	Beds	Occupation
Pirapozinho	Santa Maria Hospital in Pirapozinho	State	Private	49	59	83%
President Prudente	St. John's Hospital	State	Private	88	180	49
President Prudente	Bezerra de Menezes Spiritist Psychiatric Hospital	State	Philanthropic	116		52%
President Prudente	Allan Kardec Psycho-Physical Social Clinic	State	Philanthropic	0	60	0%

Org: GUIMARAES, Leandro (2017) Source: SES-SP and Prontuàrio, 2008.

In this context, we can consider that, unlike Foucauldian sociology, geography was a late entrant to this debate. It was only in the 1990s that the issue of mental health began to emerge in geography, with the development of important studies on the living conditions of these people. This theme took shape due to the growing diagnosis of tens of thousands of people with irreversible mental problems, as well as the growing oscillation of the human mind in the face of the characteristics that society finds itself in today (PARR; DAVIDSON, 2010).

This has resulted in the conception that the mental health of these individuals is not just an inevitable result of a biomedical condition, but also a product of socio-spatial processes that actively participate in their lives. Alves (2010) agrees with this statement, pointing out that the body goes beyond the limits imposed by the skin, having to be thought of culturally, socially, subjectively and spatially.

The broadening of the concept of health to include the expansion of life has led to new public policies being created in order to provide a new possibility for life and the construction of Places. But despite the fact that the discussion is present in geography, the subject is by no means exhausted.

In this way, the place is shown to be a vital factor, also providing a space for recovery for this social being, through the symbolic, the society in which it is permeated and the physical aspects present. TUAN (1983) highlights place as an intimate environment where our fundamental needs are considered and deserve attention. He also adds that place is not just an act to be explained in the broad structure of space, it is reality that must be clarified and understood through the horizon to which people give meaning. The relationship between man and place produces topophilic relationships.

From this perspective, Relph (1796) points out that place is inseparable from human existence, it is actions that become intersubjective, linking the individual to the world *"[...] place means much more than the geographical sense of location. It does not refer to objects and attributes of locations, but to types of experience and involvement with the world, the need for roots and security (RELPH, 1976, p. 156)".*

Parr and Davidson (2010) highlight the need for studies into the most serious cases that affect health in social and economic terms, but especially studies into the identities of these individuals and their places in community life that analyze the construction of relationships of belonging. We believe that this research can help to restructure the lives of people with some kind of mental illness by questioning the very limits of the bodies of these people with some kind of mental alteration (irreversible or not) in the places where they live (PARR; DAVIDSON, 2010).

It is in this sense, therefore, that we highlight the awareness of place as a socially constructed and complex phenomenon that has become a key point for the very understanding of health geography, establishing the relationship between place and the health-disease process and health care as an important element of analysis (KEARNS AND MOON, 2002).

Chap. 2 Autism Spectrum Disorder: Histories, Concepts and Definitions

What is Autism Spectrum Disorder?

A - Persistent deficiencies in communication and social interaction: 1. Limitation in social and emotional reciprocity; 2. Limitation in non-verbal communication behaviors used for social interaction; 3. Limitation in initiating, maintaining and understanding relationships, ranging from difficulties with adapting behavior to adjust to various social situations.

B - Restricted and repetitive patterns of behavior, interests or activities, manifested by at least two of the following aspects observed or by clinical history: 1. Repetitive and stereotyped movements in the use of objects or speech; 2. Insistence on the same things, inflexible adherence to routines or ritualistic patterns of verbal and non-verbal behavior; 3. Restricted interests that are abnormal in intensity and focus; 4. Hyper- or hyporeactive to sensory stimuli from the environment.

C - Symptoms must be present in the early stages of development. They may not be fully manifest until social demands exceed their capabilities, or they may be masked by certain learning strategies throughout life.

D - The symptoms cause clinically significant impairment in social, occupational or other important areas of the patient's current functioning.

E - These disorders are not better explained by cognitive impairment or global developmental delay. (DSM-V, 2013)

From a historical point of view, we can say that the idea of autism originated with the psychiatrist Henry Mausdley, who became interested in children with serious mental disorders in 1863, when he correlated and classified different anomalies. Until then, the criteria and terminology used to understand the subject came directly from adult psychiatry, which made it almost impossible to obtain a proper diagnosis and treatment (KANNER, 1973).

Subsequently, in 1908, Eugen Bleuler used the term autism, originating from the Greek word autos, meaning "in oneself", to report a patient diagnosed with schizophrenia, describing that he lived in his own world, as a secondary defense mechanism, protecting the lack of interaction of ideas and feelings (RITVO, 1976).

In 1928, Heller reported a pathology that was rarely found, describing behavioral withdrawal from the age of three to four (KANNER, 1973). But it was in 1943 that Leo Kanner published Autistic Disturbances of Affective Contact, describing for the first time early infantile autism, which characterized the condition of 11 children he studied, with varying degrees of the disease, which made it difficult to identify (KANNER, 1943).

These children interested Kanner because they differed from everything in the psychiatric literature. Based on his analysis, he made a distinction between schizophrenia and autism by stating that the disorder suffered by autism is not a consequence, but an initial mental failure (FERNANDES, 2009).

According to Kanner, the children had a fundamental disorder related to the inability to relate in a common way to people and situations from the beginning of their lives. The parents, when referring to the children, mentioned that they had always been *"self-sufficient"; "living as if in a shell"; "happier when left alone"; "totally absent from everything that concerned them"; "giving the impression of silent wisdom"; "failing to develop the normal amount of social awareness"; "acting almost as if under hypnosis".*

Like this,

"From the outset, an extreme autistic isolation which, whenever possible, disregards, ignores, silences anything that reaches the child from outside. Direct physical contact or movements and noises, like threats, to break the isolation, are still treated "as if they didn't exist", or, if this isn't enough, there is still the painful resentment of these children at the interference that is distressing them." (KANNER, 1943, p.16)

Almost at the same time, around the 1940s, Hans Asperger published his thesis "Autistic psychopathies during childhood", diverging from Kanner's work, especially from the perspective of language and communication. For Asperger, the autistic spectrum was a state and not a psychosis, as advocated by Kanner.[1] (FERNANDES, 2009).

Margareth Mahler, in her research on the behavioral view of children who lacked affection, published in 1952, believed in the incompatibility of mothers and children, analyzing the behaviors she called symbiotic psychosis. Mahler defines autistic psychoses as appearing early, on average from the first year of life. Consequently, these children become disoriented and also have an absence of posture, stare at nothing, absence of smiles and crises when disturbed by others, later becoming symbiotic psychoses (BRAGA-KENYON; KENYON; MIGUEL, 2002).

In 1954, Leo Kanner published in the journal Genetics and Imeritance of Integrated Neurological and Psychiatric Patterns that no organic evidence had been found to prove infantile autism. He explains that it could perhaps be an earlier form of schizophrenia, placing autism within schizophrenia in a broader sense (KANNER, 1973).

Thus, due to the extent to which the disease was discussed, children in Europe and the United States were diagnosed as autistic as if it were a fad, causing Leo Kanner to once again emphasize the uniqueness of the syndrome in 1960 (KANNER, 1973).

In 1964, Bernard Rimland published "Infantile Autism: The Syndrome and Its Implications for a Neural Theory of Behavior", considering autism as a syndrome with no connection to schizophrenia (SUPLICY, 1993).

Based on Rimland's observations, Kanner once again defended the autistic syndrome as something unique (FERNANDES, 2009). In 1968, autism took an important step forward

---

1 Hans Asparger's work only became known after the 1980s. Given the importance of his discussion of this disease, it was renamed Asparger's Syndrome (FERNANDES, 2009).

when Michael Rutter proposed four criteria for identifying it. They are:

1. Before 30 months of age
2. Abnormal social development
3. Diverted development
4. Stereotyped behaviors and routines

Later, in 1978, he launched the fifth criterion.

5. Absence of delusions, hallucinations and schizophrenic thought disorders (RUTTER, 1985).

From the 1970s onwards, Rutter and Schopler again advanced the discussion on the subject, emphasizing that autism should not be treated as a child psychosis and began to focus more on cognitive deficits, especially the lack of communication and social impairments (GADIA, TUCHMAN and ROTTA, 2004).

With these studies, we can see that over the years there has been a significant increase in research related to Autism Spectrum Disorder, both in improving behavioral and genetic treatments. In the last two decades, according to Klin and Mercadante (2006), there has been an exponential increase in the identification of subjects.

In Brazil there is also an intense debate about the disease, with various theories accepted for its classification and definition. The identification and classification method used in Brazil is that of the American Psychiatric Association (APA), which since 2013 has encompassed all disorders caused by developmental delay (Rett Syndrome, Childhood Disintegrative Disorder - Heller Syndrome, Asperger Syndrome, Autism and Atypical Autism) under a single name: Autism Spectrum Disorder (ASD). The change reflected the scientific view that all disorders are actually the same condition with different degrees, so for a patient to be diagnosed with ASD by a specialist, the symptoms must be present in early childhood, but may not manifest themselves completely until the social demands exceed the limit of their capabilities with marked abnormal characteristics in two main areas: social communication and deficits and fixed or repetitive behaviors (DSM-5, 2013).

Despite the scholar's persistent searches to find works related to the topic under study, little literature was found on the subject, which is reflected in the few studies carried out in Brazil. Despite this scarcity, there was an approximation with the works found.

A first study was carried out in the field of psychology in private schools in Porto Alegre/RS with two male students from the same class who had no physical or sensory problems and were between 3 and 4 years old. This school is considered to be small, with a total of 15 students, a nutritionist, a pedagogue and two teachers, as well as two employees responsible for the cafeteria and the general organization of the school.

The study was carried out as part of a master's thesis developed by CAMARGO, Siglia P. H. and under the supervision of BOSA, Cleonice A. (2007) at the Federal University of Rio Grande do Sul (UFRS). The two children were filmed, 30 minutes in the yard and 30 minutes in the classroom, on different days, once a week for three different weeks. As a result of the study, there was evidence of social competence in the child with autism, participating in group activities, being asked by colleagues, accepting suggestions and requests from others, being friendly, affectionate and willing. On the other hand, there were some behaviours related to aggression and disorganization of the self.

The second study was carried out with the aim of analyzing the process of inclusion of students with autism in two public schools in two municipalities in the Baixada Fluminense. The work was developed into a doctoral thesis by SERRA, Dayse C. G and supervision VILHENA, Junia (2008) in the area of Clinical Psychology. The study was supported by the Pontifical Catholic University of Rio de Janeiro (PUC-RJ). The methodology chosen was the interview as a qualitative research technique to collect data from families, teachers and students who lived with autistic classmates. The technique was chosen because the interview emphasizes the face-to-face relationship between the interviewer and the subjects of the evaluation and also seeks to make explicit the rationality of the contexts and the internal logic of the various and varied agents being interviewed. As a result of this study, the author points out that in order for children with autism to be truly included, profound changes are needed in the Brazilian education system, both in the legal and pedagogical spheres, such as teacher training and guidance for parents and guardians.

Bibliographical references

ALVES, N. C. **A cidade inserita no meu corpo: gènero e saù em Presidente Prudente-SP**. Dissertation (Master's in Geography). Faculty of Science and Technology, UNESP, Presidente Prudente. 2010.

AMARANTE, Paulo. **Mental health and psychosocial care**. Rio de Janeiro: Fiocruz, 2007.

AMERICAN PSYCHIATRIC ASSOCIATION. **Diagnostic and Statistical Manual of Mental Disorders, Fifth Edition (DSM-V)**. Arlington, VA: American Psychiatric Association, 2013

BRAGA-KENYON, P., KENYON, S. E., & MIGUEL, C. F. Applied Behavior Analysis: A Model for Special Education. In: W. Camargos (Ed.), **Invasive Developmental Disorders: 3rd Millennium**. Brasilia, DF: CORDE, 2002. p.148-154.

CAMARGO, Siglia Pimentel Hoher and BOSA, Cleonice Alves.Social competence, school inclusion and autism: a critical review of the literature. **Psychol. Soc.** [online]. 2009, vol.21, n.1, pp.65-74.

FERNANDES, F. D. M. Qualidade de vida dos cuidadores de crianças com transtorno do espectro autistico. **Rev Soc Bras Fonoaudiologia**, Sâo Paulo, v. 14, n. 3, p.6-482, agosto, 2009

FOUCAULT, Michel. **Mental Illness and Psychology**. 5ª edition. Tempo Editions Brasileiro. Rio de Janeiro. 1975

. **History of Madness in the Classical Age.** Perspectiva Publishing House. São Paulo. 1972

. **Microphysics of power**. Organized and translated by Roberto. Machado. Rio de Janeiro: Ediçôes Graal, 1979

GADIA, Carlos A.; TUCHMAN, Roberto and ROTTA, Newra T..Autism and invasive developmental disorders. **J. Pediatr.** Rio de Janeiro, 2004,p.83-94

KANNER, L. The birth of early infantile autism. **Journal of Autism and Childhood Schizophrenia**, 3(2),93, 1973.

. Autistic Disturbances of Affective Contact. **Nervous Child**, n. 2, 1943. 39

KEARNS, Robin; MOON, Graham. From medical to health geography:novelty, place and theory after a decade of change. **Progress in Human Geography**, Australia, vol. 26 n.5, p. 605-625, 2002.

KLIN, Ami; MERCADANTE, Marcos T. Autism and invasive developmental disorders. **Rev. Bras. Psiquiatr.** Sâo Paulo, 2013

MELO, Joâo Baptista Ferreira de . The triumph of Place over Space. In: MARANDOLA-JUNIOR, EDUARDO; HOLZER, Werther, OLIVEIRA, Livia. **What is the Space of Place?** Rio de Janeiro: Perspectiva,2012 pp. 33-68

OLiVEIRA, Livia de. The sense of place. In: MARANDOLA-JUNIOR, EDUARDO; HOLZER, Werther, OLIVEIRA, Livia. **What is the Space of Place?** Rio de Janeiro: Perspectiva, 2012 pp. 3-16.

PARR, Hester; DAVIDSON, Joyce. Mental and Emotional Health. In: BROWN, Tim; MCLAFFERTY, Sara; MOON Graham. **A Companion to Health and Medical Geography**. United Kingdom: Blackwell Publishing Ltd, 2010, p.243-258.

PESSOTTI, Isaias. **The century of asylums**. Sâo Paulo: Editora 34, 1996

RELPH, E. **Place and placelessness.** London: Pion, 1976.

. Reflections on the Emergence, Aspects and Essence of Place. In: MARANDOLA-JUNIOR, EDUARDO; HOLZER, Werther, OLIVEIRA, Livia. **What is the Space of Place?** Rio de Janeiro: Perspectiva,2012 pp. 17-33

RITVO E.R, ORNITZ E.M. **Autism: diagnosis, current research and management**. New York: Spectrum; 1976.

SERRA, Dayse Carla Genero. Between hope and the limit: a study on the inclusion of students with autism in regular classes. **Psychol. clin.** [online]. 2009, vol.21, n.1, pp.235-235.

TUAN, Yi-Fu. **Space and place: the perspective of experience**. Sâo Paulo: Difel, 1983

Part 2

Chap.2

What is qualitative methodology?

This procedure is based on defending the constructive and interpretive nature of knowledge, which means understanding a non-linearization of a reality that presents itself, in which such practices are inseparable from the sensitive aspects of reality (GONZALEZ REY, 2005).

Minayo (2012) understands that the raw material of qualitative research is made up of a set of three verbs: understand, interpret and dialectize, whose meanings are complemented by experiences, common sense and action.

From this perspective, knowledge is emphasized as a construction, a human production, which is not presented through general categories of knowledge, so much as an ordered reality. In qualitative research, knowledge lies in the ability to construct a confrontation between the researcher's thinking and the multiplicity of factors that coexist in each investigation (GONZALEZ REY, 2005).

Therefore, qualitative research is based on answering particular questions, in a sphere that cannot be quantified, and can in no way be reduced to processes of operationalizing variables (MINAYO, 2001).

It's important to note that the techniques chosen were designed to cover any gaps that might remain in the project. Interviews were conducted to find out the views of those responsible for the child, participant observation to analyze the child's interaction with the school environment and the doodle technique in order to understand the point of view of these subjects.

3.1 Participant Observation

Participant observation at the school was characterized as an important tool, as it made it possible to see the place beyond common sense, beyond constructed appearances, in order to search for the true essence. According to Minayo (2001),

The technique of participant observation is carried out through the researcher's direct contact with the observed phenomenon in order to obtain information about the reality of the social actors in their own contexts. The observer, as part of the observation context, establishes a face-to-face relationship with those being observed. [...] The importance of this technique lies in the fact that we can capture a variety of situations or phenomena that cannot be obtained through questions, since, observed directly in reality itself, they convey what is most imponderable and elusive in real life (MINAYO, 2001, p.60-61).

From this perspective, some steps were followed for participant observation. In the first instance, the analyses were not carried out until the researcher felt that he was part of the environment, i.e. where people's behavior was not modified by the researcher's presence. After all, the researcher is an observer who is being observed all the time (VALADARES, 2007).

The researcher's participation involved a long process. Firstly, because of the long time it took to "negotiate" his entry into the place of study and, secondly, to compile the information captured in detail (VALADARES, 2007).

In this way, the observation evolved from a phase in which the researcher made constant descriptions, obtaining an overview of the particular aspects of the social, of the interactions and of what happened in the field. With this in mind, we tried to focus on the observations of certain situations and events after analyzing the data previously collected. Finally, in order to refine the observation, we began to look for differences between specific categories that had already been identified. As Correia (2007) teaches us

Participant observation, finally, acquired a researcher/researched relationship, since the information and answers given to his questions were the result of his behavior with the group (VALADARES, 2007), and a total of ten descriptions were made on different days, not counting the days of approach to the Place.

3.2 Interview with semi-structured script.

Another research method proposed in the study, in addition to participant observation, was an interview with a semi-structured script recorded with the prior authorization of those responsible for the subjects (parents and teachers of children/students with Autism Spectrum Disorder), based on the hypotheses and theories that guide the research, based on the elaboration of a script (TRIVINOS, 1987).

As the interview unfolded, the list of questions and answers allowed the researcher to come up with new hypotheses, which had to be resolved through new questions that were still in the process of being analyzed, without necessarily being included in the script. According to Trivinos (1987), this form of interview not only favors the description of the phenomena studied, but also seeks to understand their totality.

Thus, the script assumes the form of partially guaranteeing guidance in at least part of the research, as well as being an element for the researcher to organize himself in the moment before the interview, trying to establish the questions in a precise way and with the greatest possible ease of understanding for the interviewee (MANZINI, 2001).

Based on Manzini (2001), some care should be taken during interviews. The first of these is language, since intimacy with the person being interviewed helps with the choice of vocabulary, as well as the understanding of words, which can change even within the community itself.

Another precaution was the form of the questions, using phrases that are specific, but taking care with vague terms, in an attempt to be as direct and simple as possible. (MANZINI,2001)

Finally, the interview followed a coherent and logical structure in the sequence of questions put to the interviewee. In the first instance, questions were created so that the interviewee would feel more comfortable answering them, moving from the easiest to the most difficult. We also used thematic blocks grouped in order of difficulty, according to the recommendations of Manzini (2001).

It is clear from the above that the interview required a great deal of care, from the formulation of the script to the way the words were spoken in order to create the least discomfort for the interviewee.

3.3 Scribble technique

This technique comes from the psychoanalysis developed by D.W. Winnicott in 1964 and published in 1968. His writings on the subject can be found in the book "Psychoanalytic Explorations". Winnicott (1968) points out in the first instance that this technique serves as an approach, to establish contact with the infant patient. The game is made up of two people, anyone can play, but in social life the game ceases to have any meaning. In this sense, this technique only consists of white paper and two pencils. The paper is then torn up to make it look like the game has no meaning for the child.

Winnicott (1968) then points out to the child that the game has no rules and makes a scribble, asking them if it looks like something they see or if they can turn it into something. Then the game is reversed: the child scribbles and the therapist turns it into a drawing. The therapist contributes his or her ingenuity to the drawing, as does the child, but their drawings are discarded because the child's communication/affiliation is the only one that matters as an element of analysis. Each drawing comes from an integration that is part of the person who draws it, reverberating a typically obsessive integration (WINNICOTT, 1968). It's important to note that this game is only presented as a game for the child, not asking questions that could destabilize and cause trauma.

But this technique is extremely important for the researcher who needs to know more about any traumatic relationship that may have occurred between the child and the place, since Mazzolini (2007) points out that this technique reflects on a situation, on a Place of events, in other words, a Place experienced by the child To compose the methodology of the work, the following techniques were used

data collection instruments: interviews with teachers and parents of students with Autism Spectrum Disorder and participant observation in the classroom. In order to compare the information and guide the paths to be followed during the application of the Scribble Technique, we tried to capture important aspects that were not highlighted in the discourse, because the interviews are full of people who say what they want to say. This does not mean that the research is hierarchical, but rather that the Scribble Technique is highlighted in this work. In this context, we highlight qualitative research in association with the Scribble Technique, of psychoanalytic origin, in the composition of a general methodology, which can be a methodological path that offers a contribution to interdisciplinary analysis.

The application of the Scribble Technique took place in a classroom with 28 students, of whom we would like to highlight just two (Joca and Pedro - fictitious names used in the research, students with Autism Spectrum Disorder, according to the research ethics committee, n.55005416.2.0000.5402). The technique developed by Winnicott (1968) is applied in a place where only the person who is applying it and the person who is being applied are present. However, as we are working within a school inclusion process, applying this technique in an individualized way could result in a situation of segregation between the students with ASD and the other students in the classroom, since there would be communication between them.

The process took place at the beginning of the lesson, with the help of the class teacher and a tutor who has been accompanying Joca since the beginning of the year, due to his major mental disorder. The first step was to put the desks together in pairs so that the students themselves could play the game, one scribbling and the other drawing, and then the roles were reversed. Winnicott (1968) reports that this technique can be used as a game for children, but that it would have no scientific value for the work.

However, this couldn't happen in the same way with children with ASD, since their drawings would be analyzed later. So the researcher passed on the technique's guidelines to Joca's tutor and asked her to do it with him. The researcher chose the tutor rather than the teacher to be with Joca because they were closer and could help more easily if there were any questions.

The researcher sat next to Pedro and very close to the tutor's desk, because while he was doing the activity with Pedro, he could also guide the tutor if there were any questions. The researcher felt comfortable sitting next to Pedro and perceived the same from the boy, since they had formed emotional bonds during the former's presence in the classroom four months earlier, due to participant observation.

The teacher took on the role of explaining to the class how the "game" would take place and helping to develop it while the researcher and tutor applied the technique with the students with ASD. Several students came to show us what they had drawn and were very happy with the "game". The technique was applied satisfactorily to the students with ASD, and the results expressed throughout this work.

Chap 4

Chapter summary

We understand that Space-Time, being a complex concept, makes it an instrument that is difficult to analyze, including for understanding Geography. Based on Silva's (2000) idea of subtotality, we have highlighted what we believe is essential about Space-Time. In this sense, we emphasize that it appears in two forms[a] . The first is objective and the second is subjective. Holzer (2013) considers that the body creates space-time through its movement, its action, which Santos (1996) describes as an event. In turn, these events are the result of a combination of factors highlighted by Freud (1920,1923,1927,1930), which are internal (intrinsic to the subject) and external (material reality). The production of events shapes a symbolic memory and, if this process is congruent and this relationship becomes solid, it ends up building a sense of place. However, people with mental disorders, due to the confluence of certain factors, are unable to produce a sense of Place, as shown here in Tuan's work (1983). Their relationships end up becoming more volatile, with weaker ties and very strong imaginative systems.

4.1 First approximations

Discussing this process of making sense of the world became possible through an experience of participatory observation and the scribbling technique with children with Autism Spectrum Disorder in municipal schools in Presidente Prudente - SP. The study carried out with the children required the search for theoretical references that would allow the collected material to be analyzed:

In this sense, we understand that space is the locus of the reproduction of life, not just biologically, but a reproduction of the sensations, desires and needs that are present in all human beings. It's true that without space and its atmospheric interactions in perfect conditions, we wouldn't have the diversity that is present on the planet today. Thus, adapting to space is a precondition for continuing to live in it, even though human beings have managed to transform it into their ambitions in a way that has never been seen before. But space has only become habitable for today's human presence and individuals because of its temporal transformations. That's why it's always necessary to treat space in relation to time.

Thus, for human beings, being in material space-time, in other words, governed by nature itself, has made us curious beings, firstly in trying to understand what surrounds us (where we are), creating cultural movements, socializing and, in order to adapt it to our needs, evidenced in contemporary times by advances in science, technology and information. Secondly, the possibility of understanding who we are based on our subjectification of life. I believe that the two statements are closely related (GUIMARAES, 2014).

The relationship between space and time is fundamental to understanding the object of study of geographical science. It is not possible to talk about time without space, much less space without time.

In turn, we understand that the focus of the study of the sense of place is the possibility of understanding this object - the geographical space - from the point of view of the subjects. In this way, we practice not a geography of **the** subjects, but a geography **for** the subjects, concerned with people's lives where they live, perceive and elaborate their own understanding of the world.

An important author for understanding this cognitive process of making sense of the world is Vygotsky (2000). For Vygotsky, thought and language are a developmental process that starts in childhood and can be divided into three stages. The first stage is what he called the pile - when the objects perceived and apprehended by cognition are put together without any basis. As the child develops intellectually, he or she begins to establish relationships that really exist between these objects and no longer through disorganized subjective impressions. In this second phase, characterized by complexes. The child establishes their analysis through examples. Finally, in the final stage of their intellectual development, young learners make their associations more complex through the formation of concepts.

For Vygotsky (2000 p.61),

The formation of concepts is the result of a complex activity in which all the fundamental intellectual functions take part. However, this process cannot be reduced to association, tendency, imagery, inference or determining tendencies. All these functions are indispensable, but they are not enough if we don't use the sign or the word as the means by which we direct our mental operations.

In this process of making sense of the world, still according to Vygotsky (2000), why do we create a concept? According to Vygotsky, the formation of a concept arises from the presence of a problem that is posed to the subject, usually in experiences that involve the space-time relationship.

With these reflections provoked by the literature review, we went on to study the work of David Harvey (1996), since there we were able to explore the plurality of times and spaces, as well as Space-Time as a function of the material practices and processes of reproduction of social life.

This idea was corroborated by Santos (1996). According to him, what connects social life to its material aspect are events, at a given moment qualifying them, because events create time in space, through a social subject. With each new event, past aspects change and are reconfigured into a new happening. Events, however, are the result of certain combinations of factors.

Thus, an event only happens when there is a rearrangement of factors that creates a time in space through the action of the subject, thus constituting the essence of the human being. Milton Santos (1996) advances this theoretical analysis by subdividing events into finite and infinite. Finite events are the impossibility of carrying out spatial actions at the same time. Infinite events do not have the character of competition, but of accumulation, since resources are never exhausted.

Sveda and Madajova (2012) add two types of finite events to this analysis. The first would be the restriction of a human being's capacity and needs, such as sleeping, eating, hygiene, among others. The second is related to coupling restrictions, due to not being able to stay in all places and with all people, which requires a decision between various alternatives of stay, transportation or even activity, which can be manifested as a choice or not.

Guimarães; Pickenhayn and Lima (2014) will call this process *"SigMental"*, as a process similar to the Geographic Information System (GIS), in which each piece of information is received through events, and begins to be reconfigured into Layers, which in turn are combinable with each other, which adapt us for some decision-making.

Therefore, the subjectification of space-time generates a chain of actions. Every event produces a response in space-time. When several events point in the same direction, different societies and cultures are created, which constructs the materiality in which we lose control of the action because it is already materialized. Thus, the very material built by man becomes a living organism and the fruit of social reproduction, like capitalism itself and the nation-state. However, not every response is hegemonic, and a very important aspect comes into play: the power of the response. One person can have a much greater quantitative power of response than the power of response of an entire society.

When we carry out an action and then there is a response generated, we enter another dimension, which is the extent of the response in space. We may be aware of how much power and intentionality has been projected into the space awaiting the desired response, but we will never know its real scope and impact, not only in space, but also in other geographical categories (such as landscape, place, region and territory). may seem like seconds.

In this sense, Harvey (1989) emphasizes that *"[...] every system of representation is a kind of spatialization"*, which is presented in different ways and at different times from the point of view of a child, a person with a mental disorder, oppressed minorities, women, men and different social classes (HARVEY, 1989). Thus, *"conflicts arise not only from admittedly diverse subjective appraisals, but because different objective material qualities of time and space are considered relevant to social life in different situations (HARVEY, 1989, p.190) '* the scale being the place where all these forms of manifestation and repression are registered (HARVEY, 1989).

For Armando Corrêa da Silva, 1999, p.99 *"Geography is a subtotality. It can be identified, in the sphere of knowledge, as an ideology of everyday life, expressed by the apprehension of the spatiality of the relational value contained in reality."* Thus, materiality, however complex it may be, does not explain the reasons why human beings combine certain factors to form the events that connect them with space. This is why we highlight not only the external factors in the construction of the sense of place, but also the internal factors, intrinsic to the being, which will be decisive for an event to take place in the present space-time.

Therefore, events are the result of internalized actions, i.e. past, future and unconscious times-spaces; together with the influences of external factors characteristic of human culture and capitalist society, as well as physical/chemical/biological aspects and the

State, which affect the subjects and make them produce an action.

In this sense, we understand that Space-Time relations are established by the practices and material processes of reproducing social life, marked by events that only happen from a new arrangement of factors (external and internal to the subject), which creates a temporal dimension in space through action.

4.2.3 Capital and other factors in the construction and production of space

Factors are captured by experiences through the senses, from smell, taste and touch to visual perception (TUAN, 1983). This author considers that there are *"[...] different ways in which a person knows and constructs reality (TUAN, 1983, p.2)".*

Among the factors that make us produce an action in space, I would like to highlight in this topic, capital and the advance of globalization in influencing our perception of time in space. We know from the conceptualizations of Harvey (1996), Santos (1996), and many other authors, that the accumulation of capital has always been an object of study for geography because of its spatial question, although many authors from other sciences do not emphasize this. Harvey (2004) points out that without geography as a facilitator of expansion, organization and uneven development, capitalism would have ceased to function as an economic system a long time ago.

Capitalism stands out for the complexity with which it has managed to transform other factors through human events, almost always in its favor. Capitalism has managed to embed itself so deeply in our actions that it has ended up turning our bodies into commodities, as Harvey (2004) points out. According to him, *"the worker (a person) sells his labor power (a commodity) and uses it in the work process in exchange for a cash wage that allows him to buy manufactured goods so that he can live and return to work (HARVEY, 2004, p.72).*

This process has led to a social division of labor (division of living labor and stratification into different classes) and consequent territorial division of labor (locational sense), causing different social classes to feel time in space in two combined ways, based on fixes and flows in relation to material chronological time, which is imposed on us (SANTOS, 1996). In Harvey's words,

Capital submits human forces to an enclosure within the rhythms of space and time restricted to regulatory (and sometimes spatially confined) activities, to the frequent subordination of bodily rhythms and desires (as an appendage of the machine) capitalist demands push the limits of the working body - its capacities and possibilities (HARVEY,

2004, p.147).

For Freud (1930, p. 15) the entry of capitalism as an economic system did not bring about a solution to human problems, since,

[...] seem to have observed that the newly acquired power over space and time, the subjugation of the forces of nature, the fulfillment of a longing that goes back thousands of years, has not increased the amount of pleasurable satisfaction they could expect from life and has not made them happier. Recognizing this fact, we must be content to conclude that power over nature is not the only precondition for human happiness Society and the lives of human beings have changed with the development of information and communication technologies (ICT) and their role in the daily lives of individuals has brought new phenomena that have disrupted the conventional view of human behavior in time and space (SVEDA, MADAJOVA, 2012).

However, other factors are interesting for our analysis, especially those that tend to influence us more strongly in our actions, such as cultural and social movements. Although, as I said above, capitalism manages to influence almost all events. Today, there are many counter-hegemonic cultural movements that escape capital's attempt at homogenization, such as a social/cultural demonstration against capitalist dictates. All our actions depend on the confluence of space and time, which is both constant and discontinuous, depending on the predominant factors that influence our actions.

In addition to these, other factors are interesting and necessary for the construction of events, in this sense, a first necessary classification is to understand past time. Many of our choices for holding events and providing an action in space are the result of past human times in different places and spaces, because the past shows the subjects their experiences and which they have classified into good and bad choices. Much of their past influences their present and future choices, because the past was once the present. This relationship, when introjected into the mind, establishes subjective flows and actively participates in the relationship of new spaces.

Future time, on the other hand, is not a time not yet lived spatially, it is an uncertain time, of uncertain spaces. The future shows our personal projection and what makes us want to live or die as a result of our actions in the present and the past. We imagine the future based on the present/past and change the present based on the uncertainties of the future.

Thus, Guimaràes; Pickenhayn and Lima (2014) point out that this subjectification of the speed of time is only possible through the concept of the "Extended Present", further developed by Santos (1997), which refers to *"[...] this network of changing locations of unlimited time, in which the landscape becomes visible as an intertwining of information (GUIMARAES; PICKENHAYN; LIMA, 2014.p.46).*

It's important to point out that hunger, the state, laws, a hurricane, a bomb, capital, culture, societies, among many other aspects, will be felt by the subjects from different combinations at different times and will make the subject register everything and carry out an action in space, but perhaps almost every aspect today is linked directly or indirectly to capital. From this perspective, Santos tells us that

With each new event, pre-existing things change their content and also their meaning. Events are all present. They happen at a given moment, a fraction of time that qualifies them, events are the matrix of time and space. (SANTOS, 1996 p. 145-146)

4.3 The unconscious as a factor in the construction and production of space, especially for people with mental disorders.

Based on these initial considerations, we verticalized the study of the unconscious in the process of spatial production due to the specific nature of the work *(see page 89).* In an analysis of Freud's work (1920), we consider that unconscious processes do not vary according to time. He is also emphatic when he analyzes that *"unconscious mental processes are, in themselves, 'timeless' (Freud, 1920 p.12)"*. Sandler (2010) added to his analysis that there is no such thing as mental space either, and that everything merges into the timeless and anispatial unconscious, as these are just innate characteristics of being.

Freud (1923) considers that the concepts used of the unconscious are related to powerful mental processes linked to the theory of repression and latency. For him,

For us, the repressed is the prototype of the unconscious. We realize, however, that we have two types of unconscious: one that is latent, but capable of becoming conscious, and another that is repressed and is not, in itself and without further work, capable of becoming conscious. This internal understanding (insight) of psychic dynamics cannot but affect terminology and description. The latent, which is unconscious only descriptively, not in the dynamic sense, we call preconscious; we restrict the term unconscious to the dynamically unconscious repressed, so that we now have three terms, conscious (Cs.), preconscious (Pcs.) and unconscious (Ics.), whose meaning is no longer purely descriptive (FREUD, 1923, p.13).

Based on Freud's conceptualizations (1923 p. 17), we could consider that the Unconscious would be another factor or aspect that would influence us internally in the formation of an event. In other words, the unconscious could become conscious through ideas (insights), although it itself would not become conscious. Thus,

'Being conscious' is, in the first place, a purely descriptive term, which rests on the perception of the most immediate and certain character. Experience shows that a psychic element (an idea, for example) is not, as a rule, conscious for a prolonged period of time. On the contrary, a state of consciousness is characteristically very transitory; an idea that is conscious now is no longer so a moment later, although it can become so again under certain conditions that are easily brought about.

Winnicott (1983), a psychoanalyst contemporary to Freud, proposes an advance on this theory. For him, this concept of the repressed inconsistent would only apply to neurotic people[2] , who have this relationship between the I-not-I, who recognize a material world that is external to them. However, some individuals are unable to establish this differentiation because they do not have these aspects formulated in a congruent way, which Winnicott (1983) considered to be psychotic people.[3] This incompleteness results in people without an exact delimitation of what is conscious and unconscious.

Hostile events that have taken place throughout the child's history in some part of the space can make it difficult for some people to establish a relationship of belonging with the place, and the reverse is also true. Winnicott (1990), in his work on human nature, points out that babies are born with very strong instinctive aspects, but with normal daily maternal and paternal care, the child acquires social and cultural aspects that enable them to live normally in the spatial environment.

However, in order for this to happen, Winnicott (1990) stresses that this care must be Good Enough[4] , in other words, that there must be a minimum amount of care for integration to take place, in which an internal reality is perceived that is different from an external reality, in which he begins to worry about himself and the other, not being driven by the more organic aspects such as instincts, in other words, a sum of calm and excited states.

This experience between him and a spatiality, when it happens in a sufficiently good way,

---

2 Winnicott (1983) points out that neurotic people, having reached psychic integration, still experience reflections of their past, in a question of identification.
3 For Winnicott (1983), psychosis is the result of people failing to reach the stage of psychic integration.
4 Joint family care

enables him to create a true self [5] . But until this state of integration is reached, there are failures in this process, because the environment never manages to welcome the individual in an ideal way, and these failures cause a state of repression in the child.

It is important to note that Winnicott (1982,1983,1990) proposes a different theory to that expressed by Freud regarding the theory of the conscious and unconscious. Winnicott considers that when there is support from the environment, the child leaves the instinctive aspects behind to acquire the social aspects, in which the unconscious remains only a repression of the social aspects experienced. In Freud's view (1927), the conscious is just a "protective cover", a thin film created through social/cultural rules that tries to shape the reality of the instincts that continue to exist in human beings.

On the other hand, Winnicott points out that when these environmental failures become constant, the child stops progressing mentally, resulting in a process we call false self, in which an extreme state can be implanted as a real state for the child, as a defense movement. Thus, if the false self ends up becoming organized, creating new defenses against the environment, it results in children with Autism Spectrum Disorder, who in the vast majority of cases become psychotic.

This process of near-collapse makes it very difficult for people to have a stable relationship with the spatial environments they live in, because at times when they deny spatiality and their relationship with it, they are denying themselves. This denial of a particular environment generates new negative flows that end up reaching other spaces that create new defenses with the external environment.

Finally, the study of psychoanalysis is very important for the training of geographers concerned with mental health, because spaces and places leave their marks (whether good or bad) on subjects, just as subjects leave their marks on places and spaces, even though they may be fragile, as is the case with some people with mental disorders.

Our minds have the ability to create imaginative times and spaces and consider them to be real, as if they were the result of a concrete event, often to escape the reality of material space-time. It's important to highlight this analysis, because people with Autistic Spectrum Disorder, whom we worked with in our research, because they can't differentiate, start to imaginatively create times and spaces and live within this created world. Thus, their body ends up being their only refuge. Since it's no longer interesting to live in reality, we create a space-time that meets all our needs. Some kind of capsule is created to distance you

5The self corresponds to an empirical reality that makes it possible to differentiate between the self and the world.

from the material present.

These questions bring us back to Freud (1930), especially in his book Malaise in Civilization, when he points out that we walk down different paths, and they become very different, with different priorities, either for the sake of pleasure - personal satisfaction, or to avoid displeasure. Based on these ideas, we know that many of our desires are utopian, and almost always our choices don't lead to what we really want. This is why every social being has to find their own way, their own specific way of living. Freud is emphatic when he analyzes that

All sorts of different factors will operate to direct his choice. It is a question of how much real satisfaction he can expect to derive from the external world, of how far he is driven to become independent of it, and, finally, of how much power he feels at his disposal to alter the world in order to adapt it to his desires (FREUD, 1930, p.13).

It is also important to point out that perfect satisfaction is impossible to achieve, as we will never have complete fulfillment, just like health, what we develop is an attempt at a biotic balance, always subject to dynamic readaptation in search of harmony produced in the relationship between subject-space-time and subject-place-time (GUIMARAES; PICKENHAYN; LIMA, 2014).

4.4 Understanding space-time relationships and sense of place

The literature review carried out in the research allowed us to understand that human beings need space and place. The construction of a sense of place indicates that the subject has acquired an increase in confidence, which in turn increases the need for freedom, to discover new possibilities. However, there is a confluence of another movement: when we go out in search of freedom, the fear of losing our identity with the place appears, keeping the protection of our identity alive, and the search for security again in the place.

It is in this sense that Tuan (1983) points out that Space is more abstract than Place, and that we create meanings and values as we get to know and understand it better. When these relationships become constructive, they end up giving the subject a sense of security. Place is a sequence of events that acquires meaning, but such conceptualizations take time to be consolidated.

Human beings are producers of symbols, a characteristic that belongs to our species. We can become inexorably enamored of place, even the nation state, even if this experience is not direct but limited. Places also have a core of value, which requires significant locations that attract or repel in varying degrees of nuance, depending on the interference of factors (TUAN, 1983). A very clear example offered by Tuan (1983 p.18) is that of the city, according to his premises: *"The city is a place, a center of meaning par excellence. It has many symbols. What's more, the city itself is a symbol"*, in other words, the city itself is also a place

Subjects then change their dimension of time when they move from Space to Place and from Place to Space. Tuan (1983 p.27) teaches us this when he points out that *"the sense of time affects the sense of place"*. When we leave space to enter places, it's as if the whole charge of movement contracts into a calmer state.

However, this does not mean that the sense of place is a pause in movement. The relationship between space and time never becomes static; on the contrary, places continue to receive local and global interference. This is why Tuan (1983) makes the point that it is this contact between the edges of spaces and places that makes time feel different, because if there is a pause, there is subsequently continuity. Holzer (2013) points out that the pause is not the essential condition of Place, but it runs through the tangent of the concept.

BACHELARD (1964) complements this idea by saying that sometimes we think we know about ourselves in time, when in fact what we know is a sequence of fixations in Places and non-places. In this way, symbolic memory only materializes when there is a resistance of Space (HALBWACHS, 1990), which the psyche represses at a non-objective time that can end up building Places or not destroying the subject-Place relationship (BACHELARD, 1964). For Corrêia,

"Seeing" means, first and foremost, perceiving form. Form, in its apparent form, presents itself as an aesthetic phenomenon that the eye can decipher. But "seeing" is loaded with subjectivity. Rather, the image is abstract, although time and space can be assumed in it. The colors metamorphose the real, which is then an idealized hypothesis. Overcoming this appearance means experiencing the place. The place becomes an extension of the body and mind (CORRÊIA DA SILVA, 2000, p.15).

For this reason, as Tuan (1983) points out, some mental pathologies are created, whether neurotic or psychotic, from this process, with great difficulty in establishing this passage from Place to Space and from Space to Place. Tuan (1983 p. 51) gives us an example of this process;

In open space, a person can come to have a profound sense of place; and in the solitude of a protected place, the vastness of outer space acquires an obsessive presence. A healthy individual accepts restriction and freedom, the limitation of place and the expansion of space. In contrast, the claustrophobe sees small, cramped places as oppressive, not as limited spaces where meditation or fraternal camaraderie is possible. An agoraphobe fears open spaces, which for him do not present themselves as potential fields for action or the enhancement of the self.

Carlos (2007) states that it is on the basis of these concepts that we can understand place and space, based on their process of globalization. Place is a construction of a portion of space-time, filled with multiple events and exposed to pressures at all levels. In this sense, place opens up the processes of appropriation of space. This is why Carlos (2007) states that place holds within itself (and not outside of it) its meaning and the dimensions of the movement of life, which can be grasped by memory, through the senses and the body. Place is produced in the contradictory articulation between the world that announces itself and the historical specificity of the particular.

For Carlos, from this perspective, it is extremely important to highlight the role of the body in the construction of Place. It is through corporeality that we have access to an approximation of reality, constructed spatially, in a relationship of sensation, action and desire. This opens up a study of the use of place by the body. On the other hand, it's not just the body that makes the place and its history, but also the weight of human history. After all, global monumentality in no way suppresses the more local aspects of lived spaces. According to Carlos (2007)

Thus, the analysis of place involves the idea of a construction, woven by social relationships that take place on the plane of the lived, which guarantees the constitution of a network of meanings and senses that are woven by the history and culture of civilization that produces the identity of man - place, which on the plane of the lived is linked to the known - recognized (CARLOS, 2007, p22).

Given the explanation for thinking about a social construction of the Place category, we formulated a concept based on the need to study subjects with some kind of disorder, who had difficulty establishing themselves in a place, where their relationships are more volatile, with weaker but existing links. In this sense, we highlight pivoting places and fasciculated places.

For a complete understanding of the concept, we highlight the example of trees, which was the inspiration for the concept. Trees have two types of main roots: pivoting, in which there is a main root that penetrates vertically into the ground, which is generally larger and thicker than the others, and secondary roots, which depart from it, and fasciculated, which are formed by a set of fine roots, all of which are more or less the same size and thickness, and which do not go very deep into the ground, remaining in the most superficial layers.

This is how psychotic people's attachments are: they constantly try to attach themselves to the substrate, in other words, to the place, constantly trying to maintain links, but these roots can easily be broken by some other factor that is stronger than these roots. In this sense, we should think of place as trees, roots built and fixed to the substrate of the mind, but their stem grows towards materiality, where in order to continue surviving they need aspects from outside and inside. Like roots, places create subjective flows that radiate out to other places, giving birth to new embryos.

Bibliographical references

CARLOS, Ana Fani Alessandri. **The place in/of the world**. 1st ed. Sâo Paulo: Labur Editions, 2007.

CORREIA M. da C. B. Participant observation as a research technique. Rev. Pensar Enfermagem. Vol. 13 N. 2, 2009.

FREUD, Sigmund. **The Interpretation of Dreams.** Rio de Janeiro: Imago, 1996.

. **Beyond the pleasure principle**, 1920. Rio de Janeiro: Imago, 1996. p. 11-75

. **The ego and the id**, 1923. Rio de Janeiro: Imago, 1996. p. 13-72.

. The malaise of civilization, 1930. In:. **The future of an illusion**. Rio de Janeiro: Imago, 1996.

GONZALEZ REY, Fernando Luis. Qualitative research and subjectivity: The processes of constructing information. Sâo Paulo: Pioneira Thomson Learning, 2005

HALBWACHS, Maurice. **Collective Memory**. Sâo Paulo: Vèrtice, 1990.

HARVEY, David (1989). The **Postmodern Condition.** 6ª Ed. Sâo Paulo: Loyola, 1996.

HARVEY, David. **Spaces of hope**. 3ª Edition. Sâo Paulo: Ediçôes Loyola, 2004.

MANZINI, E. J. The interview in social research. Didâtica, Sâo Paulo, v. 26/27, p. 149158, 1990/1991. MANZINI, E. J.; SIMÂO, L.M. **Forms of reasoning presented by mentally handicapped adolescents: a study through verbal interactions**. In: MANZINI, E. J. (Org.) Linguagem, cogniçâo e ensino do aluno com deficiência. Marilia: Unesp, 2001.

MAZZOLINI, Beatriz Pinheiro Machado. Scribbling to be: from self to paper. **Revista**

imaginârio (USP). Vol. 13, 2007.p.493-50.

MONTEIRO, H. R. Medicalization of school life. In: GOUVÊA, G. et al. Pesquisas em educaçâo. Rio de Janeiro: **7 Letras**, 2007. p. 68-78].

MINAYO, M. C. de S. Qualitative analysis: theory, steps and reliability. **Rev. Ciências e Saùde Coletiva,** 2012, vol. 17 n.3, pp. 621-626.

(Social Research. Theory, method and creativity. Petrópolis: Vozes, 2001.

SANTOS, Milton. The nature of space: technique, reason and emotion. Sâo Paulo : Editora Hucitec , 1996.

. The new worlds of geography. In Santos, Milton; Di Cione, Vicente. **"Geography to come. Epistemological trends"**. Buenos Aires: Cooperativa Editorial Universitaria. Tome II, p. 11-24, 1997.

SILVA, Armando Corrêa da. Appearance, Being and Form - Geography and Method. Niterói: **Geografia**. vol.2, ed.3. 2000. p.07-25.

. Appearance, being and form: Geography and method. In: **Geography: modernity and post-modernity.** Presidente Prudente, 1996

SANDLER, Paulo Cesar. On transience (ninety-two years and ten thousand kilometers later). **Cadernos de Psicanalise / Sociedade de Psicanalise da Cidade do Rio de Janeiro**. *v.26, n.29, p.87-151.2010*

SVEDA, Martin; MADAJOVA, Michala. Changing concepts of time geography in the era of information and communication Technologies. **Acta Universitatis Palackianae Olomucensis** - Geographica, Vol. 43, No. 1, 2012, pp. 15-30

SASAKI, Karen. The contribution of humanistic geography to understanding the concept of place identity. **Revista de Desenvolvimento Econômico**, vol.22, n.1, 2010, p.112-120.

TUAN, Yi-Fu. **Space and place: the perspective of experience**. Sâo Paulo: Difel, 1983

TRIVINOS, A. N. S. **Introduçao à pesquisa em ciências sociais: a pesquisa qualitativa em educaçao.** Sâo Paulo: Atlas, 1987.

VALLADARES, Licia. The ten commandments of participant observation. **Rev. bras. Ci. Soc. [online].** vol.22, n.63, pp.153-155, 2007.

VIGOTSKI, L.S. **The construction of thought and language.** Sâo Paulo: Martins Fontes, 2000.

WINNICOTT, D. W. **O ambiente e os processos de maturacao**. Porto Alegre, Artes

Médicas, 1983.

. **Human Nature**. Rio de Janeiro, Imago, 1990

. **The child and his world**. Rio de Janeiro, LTC, 1982.

Part 3

Chap 5.

What is a geographical scale?

According to Corrêia (2007), scale is characterized by three meanings. The first is dimension, such as internal or external economies of scale; the second is cartographic scale, which translates into the relationship between an object and its representation on charts and maps; and finally, used for this work, conceptual scale, which *"[....] is associated with the idea that objects and actions that are conceptualized at a given scale in which processes and configurations become specific and have their own scale of cartographic representation (CORRÊA, 2007, p.63)".)"*

In this sense, Guimarâes (2015) considers that the articulation of the scale of the body and of the nation is a fundamental link in the apprehension of this sensitive world by the place of each person. In a country like Brazil, we can consider that the scale of the body has generated conflicts, since it puts pressure on the public health budget due to instability/lack of investment. Bodily differences can serve as the basis for socio-spatial forms of exclusion and oppression, especially in cities, which have produced differentiated processes of health and illness, as well as countless movements fighting for improvements in living conditions (GUIMARAES, 2015).

From this perspective, Sposito (2011) points out that the scale of the individual on the move does not necessarily coincide with the scale of the urban space, nor can the city be understood as the sum of individuals' choices, because the essence of scalar articulation lies in conflict and not in the alliance between actions. These two planes come into antagonism as the interests of the broader scales collide with the social needs of the city scale, which in turn collide with the scales of the subjects.

School inclusion, not dissimilarly, has the same conflicting characteristics. If we take the world as an example, the processes of inclusion are different in each country, each region, each municipality, each individual, as well as the links between them. For this reason, when we take a contemporary view, we can't think of school aspects in a restricted sense, without taking the concept of inclusion and its multi-scalar approach together.

5.1 Understanding the macro scale

Understanding this broad meaning given to school inclusion, we consider that the macro scale, from a material point of view, would result in an overall picture. However, the macro scale must always be articulated with the micro scale, of the social subject, observing their experience, their day-to-day life in order to fully understand the concept.

Oliveira and Leite (2007) subsidize this analysis by considering that we must:

[...] combining the process of quantitative democratization with the process of qualifying the public school, in a political scenario that is not very favorable, since, on the one hand, there is the announcement of an inclusive political-educational proposal, on the other hand, the conditions for its operationalization have yet to be given (OLIVEIRA; LEITE, 2007, p. 512).

At the national level, although we have managed to move away from a meritocratization of Brazilian education, which excluded the poorest social classes, towards a universalization of public education, this historical process did not happen concomitantly with subjects with some kind of disability, as Giroto et al. (2014) point out, by excluding individuals considered to be outside the school's homogenizing standards. In this sense, for the author,

[...] inclusion represents a comprehensive aspect, enabling people with disabilities to have access to the socio-cultural environment, as well as to receive, in this environment, the necessary support for real inclusion in the different dimensions of life. [...] The inclusion of people with disabilities is the result of a process of reflection and awareness, both on the part of the person with a disability and on the part of the socio-cultural group, and will only come about if there is cooperation and active participation on both sides, in a process of rearranging notions and attitudes. (GIROTO et al, 2014, p. 25-26).

It is also essential to point out that, at national level, the inclusion of children and young people with Autism Spectrum Disorders was only recognized on 27 December 2012 with Law No. 12.764 Art. 3, which recognizes their rights:

I - a dignified life, physical and moral integrity, the free development of personality, safety and leisure;

II - protection against all forms of abuse and exploitation;

III - access to health actions and services, with a view to comprehensive attention to their health needs, including:

a) early diagnosis, even if not definitive;

b) multi-professional care;

c) proper nutrition and nutritional therapy;

d) medicines;

e) information to help with diagnosis and treatment;

IV - access:

a) education and vocational training;

b) housing, including sheltered housing;

c) to the job market;

d) social security and social assistance.

However, we know that there is a gap between the theorizing of laws and their practice, which led us to study the situation of school inclusion in Presidente Prudente - SP, which despite being within the state with the greatest economic power due to its industrial development and where it also concentrates the largest doctors and devices and treatments for people with some type of mental disorder resulting in an early diagnosis of the disease, according to Ipea (2011) spending on education is much lower when compared to other states.

In order to provide a methodological basis for this research, we used qualitative and quantitative methodologies by drawing up two maps, as well as interviews, participant observation and the doodle technique. With the production of the first map (figure 2), we can see, from this representation, that there is a representation of children with Autism Spectrum Disorder studying in the municipal public schools of Presidente Prudente, in some cases with more than one student, reaching a total of four in a single school. It is also important to note that these children are evenly distributed throughout the urban area, including the districts (not highlighted on the map), preferably studying close to where they live, except in a few specific cases, ranging from socially/economically vulnerable neighborhoods to neighborhoods with higher purchasing power.

Figure 2 - Students with ASD in municipal schools in Presidente Prudente - SP (2016)

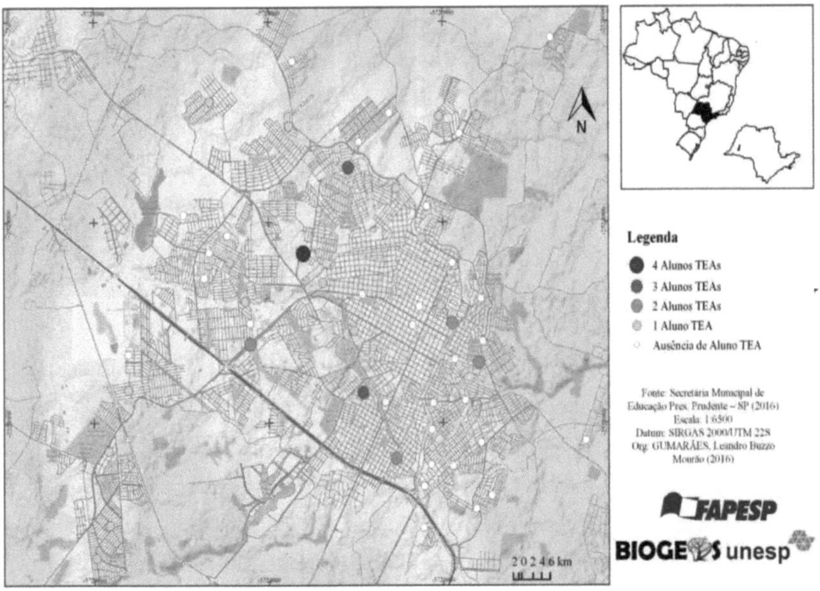

Source: Secretaria municipal de Educaçâo Presidente Prudente- SP

The second map was made on the basis of a particular **feature of** Presidente Prudente, since it has **resource rooms** in some schools - these are rooms for specialized and vertical care, focusing on the difficulties of each student with some kind of motor or cognitive alteration, at a different time to the one they normally attend. Normally, these children attend two to three times a week according to the needs identified in the students. Schools that don't have these resource rooms are transported free of charge to schools that do.

Figure 3 - Map of students with ASD and resource rooms in municipal schools in Presidente Prudente-SP (2016)

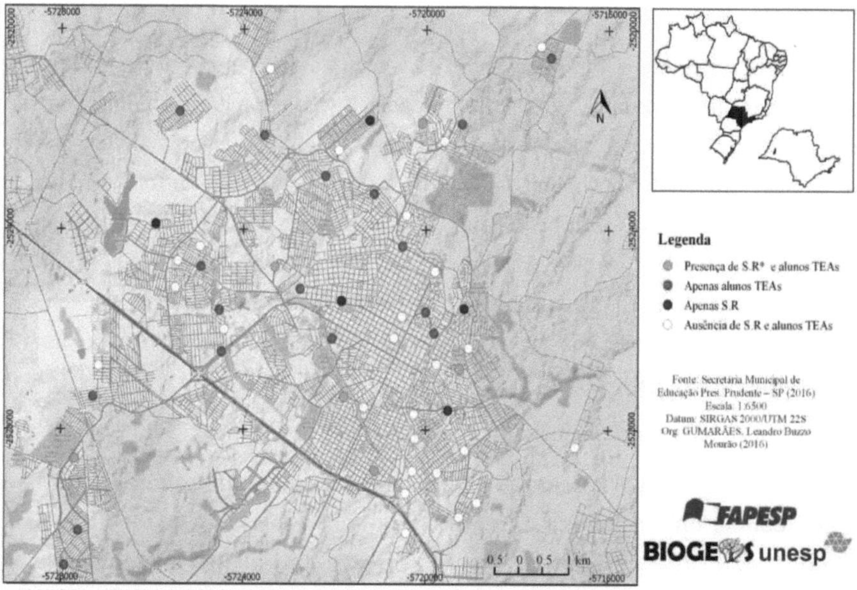

* Sala de Recurso destinada ao atendimento individualizado e especializado para crianças com algum tipo de alteração mental/motora

Source: Secretaria municipal de Educaçâo Presidente Prudente- SP

We identified from the surveys that, as there are few schools with resource rooms, they are distributed throughout the urban network in such a way that they are able to serve the other schools and not overload the others. Another point worth highlighting is the few schools that have students with ASD and resource rooms, which means that most of the children go to other schools, which can cause stress for them, since the transportation offered by the city is responsible for moving several students at the same time.

In conjunction with this spatial analysis, we have focused our studies on the production of place, which includes the subjective view of each individual who experiences the school environment, their desires and their needs. With this in mind, we have listed the main factors highlighted by the analysis of the subjects for the multi-scalar understanding of school inclusion in a school in Presidente Prudente - SP.

5.2 Understanding the micro scale

5.2.1 Resources provided by the Presidente Prudente City Hall

Presidente Prudente, according to Zoghbi et al. (2011), in their analysis of the efficiency of spending on education in São Paulo municipalities, is the administrative region that spends the most on pupils and has one of the highest results in the Basic Education

Development Index (IDEB).

This is also reflected in the maps presented, in which the number of children with ASD in schools in Presidente Prudente - SP is considerable. There are more than 30 children with Autism Spectrum Disorder present in the classroom (not counting children who are enrolled but have not yet been diagnosed), with great support for the student's development, such as food, basic sanitation and a resource room for specialized care, a room with fewer students for those with some kind of mental alteration, a tutor to monitor students in more serious conditions, dental treatment and referrals directly from schools to health units, where children receive priority care, as well as teacher training from the municipal education department.

This aspect of training was confirmed by the teacher when she stated that:

"Here at the Presidente Prudente (SP) city hall, they invited me to take part in an inclusion course on a Saturday, but on the day I had the kidney stone crisis, so I didn't go. They've already invited me twice, last year they invited me to courses, but not on special education, it was about what I worked on in the city school, traffic education". P3A When analyzing the situation in Presidente Prudente, she makes a comparison with another municipality where she worked.

"[...] in Alvares Machado (SP), I worked there for five years, we were always in contact with courses, there was the Pact, which was done with all the teachers during HTPC time, letramento, which was a course run by the city hall there" P3A

Another fact confirmed by the oral testimonies was the presence of free transportation offered by the city, which according to the grandmother of the student with ASD, the transportation offered by Seduc managed to reduce the time spent caring directly for her grandson.

"Yeah, because Pedro was getting too stressed out, I was leaving here, I would take the bus to the center and then from the center I would take another one there, because Joao doesn't pay, he has a card, those blue, pink cards ....I got for him, so he would take two buses and arrive already tired and stressed at school. Ai.... The school secretary said: - Why don't you ask... try Seduc for transportation for him? So she said during the adaptation week when school started. Then she decided to call Seduc and asked what she needed. Then she arranged the documents herself. Ai.... She told me to take them to Seduc" AP

On some days when the school was only open part-time, especially during the first week of school (the adaptation phase), Grandma would wait at the school to avoid having to get home and have to return to pick up her grandson from school, since she used public transport. According to her,

"[...] Because during the week of adaptation, it was from seven until nine, and I was there waiting, because by the time I got home, it was time to go back because the bus takes so long" AP

However, despite having obtained free transportation for his grandson from the Municipal Department of Education, Pedro, like so many other children who use this transportation, still spends a lot of time in these vehicles, which, specifically in the case of children with Autism Spectrum Disorder, is an aggravating factor, since they are highly irritable from traffic noise. According to his grandmother's accounts of the time he leaves the house and the time he arrives at school, it was found that Pedro spends around 40 minutes on the journey home/school, a relatively long time given the size of the urban network of Presidente Prudente (SP).

Like this,

"So, he comes here at half past six, then he goes back. He has a resource room today, which was yesterday and changed to today. So... he comes for two hours to pick up Joao and then he brings him back" AP

In this sense, when we compare the commute with public transport (common) and that offered by the Municipal Secretary of Education, we can see that the commuting time differs little from the home-school commute.

5.2.2 Structure of the school and its surroundings

All the information about the study area was taken from the school's political-pedagogical plan drawn up in 2016, and the result of observations of the place made by the scholarship holder. The school's surroundings are characterized by families from the most diverse social backgrounds, as well as brick houses next to wooden houses, including very humble ones.

The neighborhood has a public health unit, electricity, water, sewage, telephone and internet access. It is served by several bus routes. It has bars, markets, a butcher's, a pharmacy. It has a public leisure area with a sand volleyball court, sand soccer, a children's playground and a walking track. There are other school units such as nurseries and state schools offering primary and secondary education.

It is also close to a Cultural Center, which offers free access to culture, such as book loans, an amphitheater, community rooms and access to social networks and research. Social problems such as unemployment, alcoholism, drugs, promiscuity, lack of family structure and others directly affect the residents.

Therefore, this school is inserted in a community with diverse characteristics and needs to work on the problems of this society, trying to solve or alleviate them through reflection, dialogue, analyzing them together with its students and school community, to create strategies for survival and even coexistence, in the face of this situation in which we find ourselves.

However, the students attended come not only from this neighborhood, but also from neighboring neighborhoods and even far away, since the local community is aging. Many students have grandparents who live nearby and come with school transport.

Having a diverse range of students brings varied customs and behaviors from their family and community culture which, when involved with each other on a daily basis, leads to positive and negative exchanges, such as fights, disaffection, in short, behavioral problems related to social conduct that need to be resolved and investigated in order to be reprimanded in good time. To do this, we need competent and committed professionals and an open school with a participative and collaborative management. The school unit provides the following types of education: Early Childhood Education - Pre I and II (from 04 to 05 years old), Primary Education, from 1st to 5th Grade (from 06 to 10 years old) and Specialized Educational Assistance (AEE). Early Childhood Education is concentrated in one Pre I and II class, so the classroom is multi-serialized. For some years now, the demand for pupils in this neighborhood has not been sufficient to open a classroom for each class. For reasons of internal organization, the classroom is open in the afternoon. The 1st and 2nd grade classrooms also work in the afternoon, making a total of 4 elementary school classrooms and one kindergarten classroom at this time.

The 3rd, 4th and 5th grades work in the morning, and this division has helped the teachers to organize their subjects and get along with their peers. There are therefore five elementary school classrooms in the morning.

The Specialized Educational Service has two classrooms, one in the morning and one in the afternoon, to serve students at the same time as the regular classes. Assistance is provided in small groups or on an individual basis, depending on the student's disability or the proximity of the characteristics of the difficulties to be addressed.

Students are grouped in classes made up of children of both sexes and organized according to age group, with some exceptions for students who may be age/grade delayed due to attendance and/or learning problems.

The Comprehensive Education Program caters for elementary school students who want to join up to 120 places. The classes are divided into four, two in the morning and two in the afternoon. The criteria for organizing the classes are age / grade and learning level.

The school building consists of a first floor and a first floor, with the regular classrooms located upstairs. In 2014, maintenance work was carried out to improve and adapt access for the disabled, and the school now has a ramp in the library with handrails and adapted toilets. The library has ample space to serve the school and local community. It has a retrained teacher who organizes, shelves and lends out books and DVDs. It currently has approximately 3 encyclopedias for consultation and research; 50 up-to-date dictionaries; 500 children's books; 30 children's books (pre-school) and several books of Brazilian literature for adults. TV Escola and children's story DVDs.

Table 2. Characterization of the school's physical space

Dependencies	N°	Furniture	Teaching materials and equipment	Conditions of use	Type of accessibility.
Board of Directors	01	01 computer desk, 01 chair and 01 office table, 02 cupboards, 01 fridge and 02 upholstered chairs.	Internet access; 01 microcomputer, 01 printer and 01 ceiling fan, 01 camera	Adequate	Barrier-free ground floor.
Salad Community Teacher	01	01 table, two chairs, 02 armchairs, 01 steel cupboard.	Internet access; 01 microcomputer	Adequate	Barrier-free ground floor
Secretariat	01	02tables of office, 03 swivel chairs, 01 steel cabinet, 12 filing cabinets, 01 balcony and 02 small tables.	Internet access; 02 microcomputers, 02 printers, 02 telephone handsets, 01 modem and 01 ceiling fan.	Adequate	Barrier-free ground floor
Salade teachers	01	02tableswith chairs, 01 coffee table, 01 steel cabinet	Internet access; 01 ceiling fan; 01 computer; 01 printer.	Adequate	Barrier-free ground floor

		cabinet, 01 refrigerator, 05 armchairs.			
Salade Educational guidance	01	01 table office, 01 chair, 02 armchairs, 01 typist's desk, 01 keypad , 02 cabinets steel, 01 shelf.	Internet access; 01 microcomputer, 01 printer and 01 ceiling fan.	Adequate	Barrier-free ground floor
Salade Resources	01	02 shelves of steel filing cabinet, 01 steel cupboard, 01 table with 06 chairs, 03 tables for microphones, 03 chairs typist, 02 chairs office, 01 desk.	Access à Internet; Books, cds, dvds, toys, games pedagogical; 02 microcomputers, 01 printer, 01 scanner,	Adequate	Barrier-free ground floor; Floor rubberized.
Reading room or library	01	12 shelves, 10 shelves, 01 large table with chairs, 01 computer table; 02 chairs, 04 sets of tables (honeycomb), 05 steel filing cabinets, stretchers with 10 seats.	Access à Internet; Textbooks textbooksand paradidàticos, magazines, maps, globe, toys, educational games; 01 computer, 02 fans.	Adequate	Ramp handrails
TV and video room	01	30 chairs, 01 rack	Access à Internet; DVDs;01 29" TV, 01 home teach, 01 Datashow, 01 screen, 01 air conditioner and 01 ceiling fan.	Adequate	Barrier-free ground floor
Salade	01	01 table	Internet access; 17	Adequate	Tactile flooring

IT		office, 16 microcomputer tables	microcomputers and 01 printer, air conditioning.		up to the stairs. Stairs e elevator
Classroom	13	01 table teacher, 01 chairteacher , games and chairs student, steel cabinets.	Ceiling fan and mobile.	Adequate	Tiled floor up to the stairs. Staircase elevator
Games and toys room	01	04 tables with 04 chairs each; 01 teacher's table with chair, steel cabinet, 03 shelves.	Educational toys, games, puppets, educational toys, dolls and strollers, pans, etc. 02 ceiling fans.	Adequate	Tiled floor up to the stairs. Staircase elevator
Salade Nutrition	01	01 table with chairs, cupboard and pans	Utensils of kitchen; 01 fridge, 01 stove.	Adequate	Barrier-free ground floor
Warehouse	01	02 steel cupboards, 01 table office, 01 chair , steel files.	Various teaching materials; 01 television set	Suitable	Barrier-free ground floor
Deposit material Cleaning	01	03 shelves		Suitable	Tiled floor up to the stairs. Stairs e elevator
Pantry	01	Fixed shelves e table	02 freezer, 01 refrigerator, 01 ceiling fan.	Adequate	Ground floor
Cafeteria	01	10Tables , 20 benches, sideboard.	Food and kitchen utensils ;01 Thermal sideboard for serving meals, 03 ceiling fans.	Suitable	Tile flooring
Covered patio	01	01 bancode large wood.		Suitable	Tile flooring

Quadrade sports discovery	01	Soccer goalposts	Various teaching materials for sports.	Suitable, but the crossbar needs to be put in place.	Ramps e handrails
Quadrade indoor sports	01	Soccer goalposts.	Various teaching materials for sports.	Adequate, but the crossbar needs to be put in place and the tables removed.	Ramps e handrails
Internal circulation	03	Drinking fountains	Specific place for brushing and for drinking water; Freezer (for the students' drinking fountain).	Adequate, just need to finish installing the accessible drinking fountain	Tile flooring
Kitchen	01	01 table with 02 chairs, counter.	01 industrial stove, 01 blender industrial, 01 comume blender 01 ceiling fan.	Adequate	Tile flooring
Service area	01	01 table with chairs, 01 shelf, 02 cabinets, tàbuadepass clothes.	01 microwave, 01 stove, 01 tank washing machine, 01 washing machine, 01 ceiling fan , iron.	Suitable	Tile flooring
Bathroom of employees	02	Sinks and toilets	Door Paper e soap dish mirrors.	Suitable	Ground floor.
Bathroom of students	02	Sinks and toilets	Soap dish mirrors.	Suitable	Tiled floor to the vicinity. Ground floor.
Bathroom with accessibility	02	Toilet with support bar	Soap dish mirrors.	Suitable	Tiled floor to the vicinity. Ground floor.

SOURCE: Political and pedagogical project

Org: GUIMARAES, Leandro Buzzo Mourâo (2016)

In terms of staff, the school has a pedagogical coordinator, a clerk, a dentist, fourteen teachers who work in various specialties within the school, four people responsible for general services, a principal, a library assistant, an inspector, three cooks, an administrative assistant, an assistant for the games room and computer room, three caretakers, a janitor, and a trainee from the city school.[6]

5.2.3 - Training for teachers and tutors

According to the analysis based on the interviews and participant observations, we can still see a lack of preparation in dealing with school inclusion, not only for children with ASD, but for all students with some kind of disability, as recognized by the statements of both those responsible for these children and the teachers. Thus,

"It's... a lot of the time when we ask for help, people don't know how to do it, they don't know how to deal with it and a lot of the time we have to seek help outside the school to try and solve problems, right?" AP

According to her grandmother, when analyzing why professionals in the field are not prepared, she gives us an example:

"So much so that the teacher, on the day of the meeting, said that the other day in the playroom, I don't know what happened, he started crying and throwing tantrums and doing, I don't know what! She didn't know how to deal with it and he complained to me that he thought they had a day to play in the playroom, he came in saying that, that the teacher took him out of the room and wouldn't let him play!" AP

Compared to another school where Pedro studied, she reports cases of physical aggression, verbalizing that they had squeezed her arm.

[...] Then he was four years old. When I went to take him to nursery school, I was the one who took him. Then I spoke to the principal and told her. - Oh, Pedro said he'd been hit on the arm. She said, "No, we just held him like this!" (Mum simulated how the principal said that the teacher held him). [...] Tah, then I went to see, that he really didn't lie, that they squeezed his arm, because those fingers were just right, oh!" AP

From the researcher's observations, this fact can be seen within the school environment

6 Comprehensive Education Program of the Government of Presidente Prudente - SP

and at many times teachers, and especially tutors, don't know how to deal with children with some kind of disability, because many new situations arise within the school environment, where these professionals can't keep up with the same progression. Of course, we have to bear in mind that we are dealing with teachers first and foremost, not doctors/psychologists, who can diagnose and treat each child's problems on a need-to-know basis, since the specialists in this field of knowledge themselves don't do it perfectly.

Such diversity was reported by the teacher when she pointed out that:

"My class at university was very good, there were girls who already worked and knew the day-to-day life of children, so whenever there were classes to talk about day-to-day life, it was very controversial and they brought a lot of experiences... And then we debated a lot, so... I knew a little bit about reality from the work I did, which wasn't day-to-day, directly in the classroom, but what the girls brought was very rich, so it was very good. What I realized was a little lacking when I started working was

Methodologies, you need more instruction on methodologies, because when I started teaching all my classes had special children, but my first class was more special (in the sense of having many students with greater difficulties), it was a class that had been formed from failed students, and that class of children who couldn't learn. So I taught in the morning and in the afternoon I would go to Regente (Regente Feijó - SP, the town where the teacher lives), to the school to watch lessons, to see how the teachers worked so I could take them there (to the school where the teacher taught)" AP

Secondly, the diversity of students who appear in the school environment makes it difficult for teachers to adapt a methodology that takes into account the reality of the school; in order to work more vertically with the student, it is necessary to have a tutor who specializes in working with the needs of each child.

In this context, according to the teacher, the tutor provided enormous help in dealing with the student Joca. See the account below:

Wow! She helps a lot! Ideally, I wouldn't want to leave her next to Joca (an ASD student) so she could do what you do (researcher), walk around the room, helping other children, so he (Joca) doesn't feel dependent on her, but as she's having a lot of problems with aggression, she needs to. All the proposals, everything I ask for, she's always available. She's a good tutor, and I like her a lot. P3A

However, in living with the reality of the school, it was possible to see how unprepared another tutor was, who established a method of teaching through shouting and fear, so that the child was unable to grasp the information and consequently failed to carry out the activities. As has already been pointed out, this fluctuation in the way these pupils are treated means that there is a real need for specialization in caring for these children. Some teachers who taught these children in later years in the school environment panicked when the subject came up, because they didn't know how to deal with these children in the future, even causing tension in extra-class meetings about which class these children would go to.

In providing conditions for all people with some kind of disability/alteration to enroll in public school, the same conditions were not provided for the specialization of teachers and tutors. Without methodologies to care for these children, the hospital culture was transferred to the school environment. More often than not, teachers were given the task of giving medication to each student during class time, the role of diagnosing the student, and the need for medicalization to solve problems, in which many families even started to leave these children at school, so that the teachers could really offer hospital care, so that the parents could work, or carry out any other activity.

It's these aspects that allow us to point out the contradiction between inclusion, which takes place in practice, and the role/needs of the school and the teacher, with the rare exception of theoretical aspects. In order to do this, we need to reformulate the role of the school to the practical needs of each place (as far as possible), because each place has its own objective and subjective relationships, which in turn have characteristics that produce them. But how can this be done? In order for this to happen, the school needs to change its vertical vision to a horizontal one, the community needs to decide what the school needs by actively participating and establishing counter-hegemonic movements that still bind the school, such as the strong presence of Christianity in school environments, parents or guardians need to get away from the idea that the school represents a burden of inconvenience such as meetings, and above all that there is no idea that the school for people with disabilities is not a place perfected from the old asylum/hospice systems. It is important to emphasize that being included in school does not necessarily mean being included. This requires a relationship of belonging to the place, and much more, a construction/production of this social place.

5.3.4 The school's cultural context

Cultural practices within the school are still very strong. The predominant use of blackboards and chalk to apply activities and exercises is one example. Recently, students are gradually having more contact with computers in school, but still in a very superficial way, only through the computer rooms.

The school also hosts festivals and commemorative days, such as Junina, Folklore, Indian Day, which allow these children to establish a relationship with their past, and their cultural identities as Brazilian citizens.

On the one hand, it is necessary to affirm the importance of some cultural practices that are present within the school environment. On the other hand, some practices need counter-hegemonic movements, such as the example of the strong presence of Christianity, still employing beliefs and values, within the classroom, whether spread by the teachers themselves or by the school institution, which forgets its role as a secular institution.

Looking at this cultural perspective, which can eventually cause problems for schools, is the constant need to medicalize students. Its cultural framework has already been pointed out in the breakdown above, but it deserves more in-depth study because of the dimension it engenders in the school context.

Often within the teachers' council, the use of medication for these students was discussed almost as an imposition. In this particular case, the use of the medication was not necessarily to improve the student's performance, but rather as a way of calming him down, so much so that the teachers already signaled when the student hadn't taken his medication, because when he took it, he became calmer, If children with some kind of disability presented themselves differently and constantly over a long period of time, the problem for many of the teachers would be easily solved by increasing the dose of the medication, leaving the historical aspects of the children for a secondary analysis, also due to the absence or lack of collaboration between the school and the families of these subjects.

However, this masks the real negative effect that this constant change and increase in dosage can have on a child, as reported by **AP:**

"Look! He was taking Ritalin twice a day, Daforin at night and Ariprazole, then the doctor cut out the Daforin and put him on Depacote, but I? When he started taking Daforin he was a bit strong, when he started taking Ritalin he stopped eating, he started getting thin, thin, thin, thin, but now with the change of medication, when he finishes dinner he already says he's hungry. AP

The use of medication is so widespread and constantly changing for Pedro that even **P3A is** confused as to which medicine her student is using. This shows how predominant the cultural factors were in this study, now no longer because of the chemical action of the medicine, but because of the culture of medicalization in society, in other words, even though the teacher doesn't like the use of the medicine and realizes the effects caused by the child, she still thinks it's necessary.

Pedro, I think it's the third medication, he used to take Ritalin, Depacote and Aripriparazol and the Ritalin, I don't know if the Ritalin was taken away, because it was causing insomnia, causing hallucinations, sleep disorders, so he changed... He's taking stronger drugs and he's still behaving in a very complicated way. He told us to talk to his psychiatrist, the psychiatrist is going to give us an appointment to talk about it, it's very strong drugs, it's a bit more limited. So, in short, it's necessary for these children, for the time being, but it's something I don't like. P3A

This analysis also raises another issue, in the contradiction of the statements expressed by **P3A**, who points out that Pedro's behavior is still complicated, even with the use of medication and the need to use it in early childhood. However, she later mentions that she doesn't like the use of medication, which shows us that she is beginning to become aware of the real situation that these synthetic formulas can bring to these children. In the interviewee's words,

I don't like medication very much, there are many children who have learning difficulties, and they use Ritalin and other medication, but they don't need medication, they just need attention, to learn to concentrate? But in the case of these two children, it's necessary, because when they're not medicated it's much worse. P3A

Moving from this perspective to the theoretical field, Monteiro (2007) refers to the process of not learning, or the difficulty of learning, as a disease, affecting three main areas: inattentive, disobedient and hyperactive, coincidentally, as we can easily identify these same factors as being the main reasons for teachers complaining about their students. Ritalin, for example, has become the "cure" of the century for these students at school.

The schools we have observed have also strongly demonstrated a culture of preventing problems. How do you prevent aggression? How to prevent tooth decay? How to prevent students from being undernourished? The rare exceptions that escape these aspects and point to a paradigm shift towards health promotion practices are still conditioned to a change in the behavior of these individuals, as in the analyses by Lara-Silva and Sena

(2010).

5.3.5 Interactions in the school environment: from the classroom to the community

The school has a unique configuration which should be highlighted in order to understand the subject of the observation. In view of the large number of students who are not yet literate (or semi-literate), the coordinators, together with teachers (3A) and (3B), exchange students.

This system consists of leaving the pupils with difficulties in the 3A classroom, where the teacher works more intensively on the literacy process, and in 3B the pupils who are considered to be age-appropriate are left for the first shift. In the second shift (10 a.m. to 12:10 p.m.), the children return to their respective classrooms and the teachers work on the didactic content proposed for the third year.

This explanation is necessary because in the 3A classroom there are two students diagnosed with ASD, but one of them, Joca, was considered to have greater difficulty with literacy and stays with the 3A teacher in the first shift, while the other student with ASD - Pedro - has a corresponding level of literacy and stays in the first shift with the 3B teacher. This made it necessary for the fellow to interlocute between the two classrooms in order to observe the two children in their relationship with the place. On this subject, **P3A** said:

It helped a little, because the children who weren't literate had a leap forward and others had the opportunity to work on what they needed, right! Text production, punctuation, but the other teacher's classroom was very crowded and, in a way, she couldn't do this work, not because of her, because of the conditions, while I had few and many dispersed, she had a very crowded classroom. So... it helped, but if it were possible to have three classes instead, with fewer students in each class, it would be better for everyone! P3A

The social interaction between the children with ASD and the other pupils at the school took place spontaneously, by the pupils themselves, through games and the difficulties they faced, for example, when Pedro threw himself on the floor and started crying because he had lost at soccer, some of his friends stood nearby and started saying encouraging words that made him stop crying and get up. Another interesting event was when Joca managed to write a syllable of a word and the whole class, including the teacher, celebrated together.

Arguments with other classmates often came from the students with ASD themselves. A very clear example was when Pedro started saying that a classmate had reverberated offensive words, saying that he was stupid, that nobody liked him and therefore he wanted to leave school and kill himself, yet the classmate hadn't said anything. This process happened several times, like an illusory process.

Pedro's grandmother confirms his behavior;

"Look, it's... He doesn't complain! He... doesn't complain about school! It's those little complaints of his that he says, like he used to say, that he was bullied, now this year he's saying a lot that the teacher is shouting at him.

And... he says he plays float. He doesn't complain, though! It's those little complaints that come from his own head" AP

When addressing this behavior of students with ASD, the teacher points out that;

"This year Joca is a little calmer, but he was also like Pedro, very agitated and aggressive, but Joca has adapted to the medicine, he's calmer now. Joca resisted a lot at first, he resisted even more than Pedro to do the activities, but when I put Vini on his side, it was like that.... I didn't think it would work so well, but the dispute between the two of them, saying that one was finishing before the other, that was positive and Joca is starting to read, he's putting the syllables together, starting to develop this part of written reading" P3A

"They both have a degree of autism that you can't anticipate, the right thing for us to do is to anticipate so that they don't explode (he makes a sign with his hands to explode), but it's difficult to know when to anticipate, because as I said, Pedro sometimes plays (in the behavioral sense), and Joca explodes when Pedro plays, and then you have to do everything so that Pedro doesn't explode. But when they come back from the break, they're back.... Because Pedro wants to play soccer, and then the boys don't want to play with them anymore, because he's a bit uncoordinated, so you put him in goal and he doesn't want to stay in goal" P3A

The 3A teacher contributed to the students' social interaction by choosing the chairs they were going to sit on, always trying to move the students around in different seating arrangements and with plenty of methodologies for student interaction. This process didn't happen in the same way with the 3B teacher, where the desks were always separated and the students rarely changed places.

Reporting on her classroom, P3A says;

[Very agitated, noisy, but they're very smart, very intelligent, they question a lot,

they ask questions, they like to learn, but they need a bit more restraint to know when to ask, when to ask their questions, because they want to talk all the time. And, I see my classroom very agitated and with a lot of difference in learning levels. We know that it's important to work with heterogeneity, but when it's too far apart, you get a bit lost, not knowing what to do, which group to attend to exactly, if you're giving more attention to one than to another." P3A

Social interaction between students with ASD and the rest of the staff is peaceful, except when they get very nervous, which is not uncommon. When this happens, the family is asked to come to the school to talk to the teacher and the pedagogical coordinator, but the relationship with the school is sometimes not always friendly due to conflicting ideas about the child. The relationship between the school and the rest of the community happens when there are themed parties and when the family is asked to take part in student meetings.

The 3A teacher comments on her relationship with Joca's family;

[...] I have to show you later a recording I made that was presented in a case study of his at the HTPC, we did a study of his case and proposed activities, then I asked the management, then as I said, if we ask they offer to help... Joca is such a child... his family is very complicated. He lives with his grandmother and grandfather, and they both look after a girl, a girl who is in a vegetative state, so as well as the girl there's Joca who has autism. Joca's hand also has a disability, but I can't say which one, and his father is a bit absent, from what Joca tells me, he doesn't talk, he starts to... hitting. The parents and grandparents are neighbors and there are no other children nearby, he plays at school and at home he plays video games. His social problems mean that we have to invest in this, but on the contrary, he's separated from it, he doesn't have it. And the school ends up being in charge of functions that would also be maternal and paternal. P3A

Figure 4 - Map of the arrangement of students in the classroom and class tidy up

5.3.5 Historical and social context of the subjects

Pedro

The researcher's drawings were removed from the analysis because they had no relevance to the research. Before starting to apply the Scribble technique, I tore the sheet in half and then made a scribble. The drawing made by the child with ASD was of a duck alone in a pond. This drawing of the duck was repeated in his third characterization, reaffirming his loneliness. During the participant observations carried out in the classroom, Pedro constantly reported that nobody liked him, that his classmates were talking dirty to him, that he wanted to kill himself because he was the problem.

Figure 5- Duck in the Lagoon Figure 6 - Duck in the Lagoon

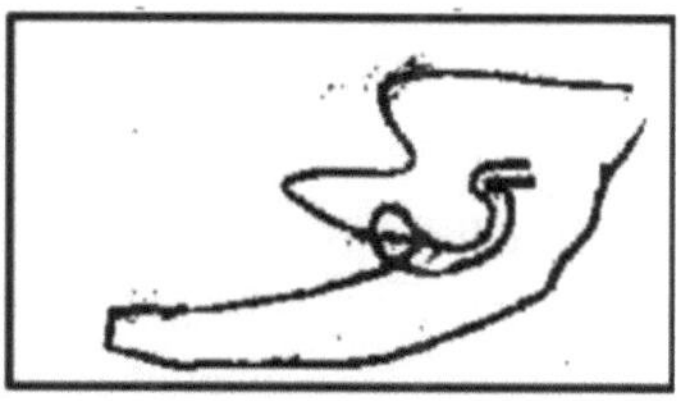

Source: Pedro Source: Pedro

We continued our game, where I made another doodle and the resulting drawing was another duck, only this time, the duck and mother were facing away from each other. During the drawing, Pedro said that it was him and his mother. This drawing proved to be very pertinent, given the specific case that Pedro had gone through in his life, because his mother had never bothered to take Pedro in. Her pregnancy was complicated, she did the lessons at school instead of helping, she didn't like to give him a bath, she gave him very little attention, she had no patience, he went to school when he was 1.5 years old, he had a very difficult time adapting, he cried, he had a fever.

Pedro, according to his grandmother, who got custody of the boy, confirms this very complicated and painful life story and his mother's abandonment. According to her details;

"The story of the mother and father, we don't know who the father is because she... [Pedro's mother], doesn't have anyone to make her say who the father is. [Pedro's mother], there's no one to make her say who the father is, now, the mother, when she was living here with us, she and Joao, they didn't get on, and she wasn't a loving mother to him, she wasn't [indication of certainty], she didn't pay attention to him, she fought more than she paid attention, there was very little attention at all" AP

He feels isolated from other people, especially his mother. The symptoms of this illness clearly appeared when he was four years old and worsened when his mother went to Dubai, in the United Arab Emirates, where he was still very young and already had stereotyped head movements and vision problems. This abandonment is projected onto his construction of Place, in which he finds it very difficult to establish relationships of belonging, since he can be abandoned again, transposing from his past to his present, as a defense movement in a constant attack by the false self to cover up the true self, since hiding the true self hides his reality. Still according to Grandma's accounts;

He shook his his head, so much so that he went [pause].

When I took him to the ophthalmologist, he went to a specialist, the doctor did several tests on him and it turned out that he had a sight problem, so the specialist said: - look, if the eye test had been done, as it has to be done in hospital, John would have been wearing glasses. But it's... this problem with his autism, it was the psychologist at the... CI, Dr. Ana. Then she referred him to Dr. Armenio, the neuro I took him to before, she also suspected it, she wasn't sure, she gave him the medication, but she suspected it, and it was Dr. Armenio who confirmed his autism. AP

Figure 7 - Paw and her cub in the pond

Source: Pedro

The next drawings made by Pedro helped us to better understand his current social context and what he goes through in his day-to-day life living with ASD, a drawing depicting a bird, but when this bird tried to fly, there was always a storm and constant lightning strikes. When we analyze this drawing, we see that Pedro is the animal, in other words, he was trying to grow and develop, but something always happened to get in the way.

Figure 8 - Airplane flying with lightning

Source: Pedro

This drawing in figure 9 complements the previous one, which, according to Pedro, represents a certain person who created a robot dinosaur and whose purpose was to carry a letter. The student did not reveal the content of the letter to the researcher. When he showed the drawing to the teacher, she said that Pedro only communicated with his mother through letters, as she lives in Dubai, United Arab Emirates. In this sense, the drawing could relate to his need to speak and communicate with his mother.

Figure 9 - Robot dinosaur

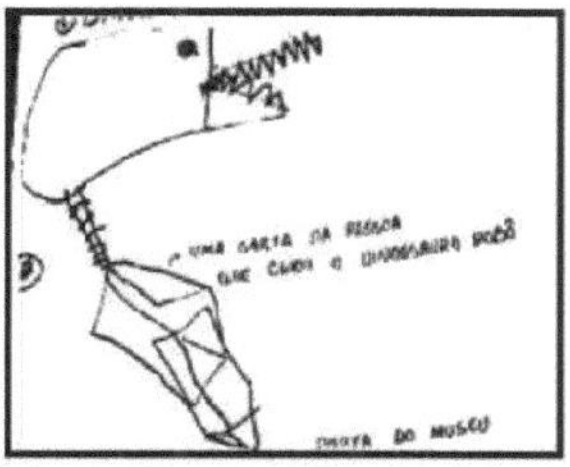

Source: Pedro

After in-depth studies, we concluded that, as the boy is unable to continue developing, he creates a false self, expressed by the bird that is able to continue its journey because of the storm, and then appears in a second moment transformed into something stronger, a robot dinosaur (probably due to the influence of the games), as the bird was not enough to hide his difficulties, in this sense, the robot is more resistant, built with materials that are more rigid to impacts, in addition to its enormous size with large teeth. However, it should be noted that the dinosaur has problems with its physical structure, it seems a little disproportionate, missing parts (lack of integration), it doesn't have a very defined body, the "arms" are kind of joined or fused to the body (like something undeveloped) and the "legs" don't appear. This boy dressed as a dinosaur robot shows aggression in his jaw in a body that is somewhat weakened.

In figure 10, Pedro showed almost a collapse of his psyche, reporting that the museum door was closed, as well as another drawing depicting a bomb and that it was on the 99th floor just a moment away from exploding, and suddenly the bomb exploded. We can clearly see a plea for help since the bomb inside him is already disintegrating, he's tired of trying to fly and falling repeatedly, reverberating even in the desire to kill himself, as he comments in the school environment.

Figure 10 - Museum door Figure 11 - Bomb

Source: Pedro Source: Pedro

Joca

Joca's case became very complicated for analysis due to other aggravating factors such as mental retardation, in addition to Autism Spectrum Disorder, so the first result expressed is that this technique is not very efficient with people with major mental disorders, given the possible flaws in imagination and symbolic capacity (TAMANAHA, 2006). However, when someone is present to carry out a directed activity, as in the case of the doodle game, a certain exploration of the activity is possible, even if it takes place by imitating the researcher's act, which in this case is drawing. In this way, much without form and characterization, we are able to extract some information. In this context, four drawings stand out.

The drawings are precarious due to the impairment of the disease, associated with intellectual disability. Difficulties with symbolic representation and creativity leave the drawings poor, almost formless, or even missing parts, but from the boy's verbalizations and the specific characteristics of his drawings it was possible to learn a little more about him.

From the participant observation and interviews, we can understand that Joca lives with his grandparents, but his mother still lives next door. In addition to caring for him, his grandparents also look after another granddaughter who is in a vegetative state as a result of a gunshot wound by the girl's own father.

Figure 12 - Shark Figure 13 - Shark

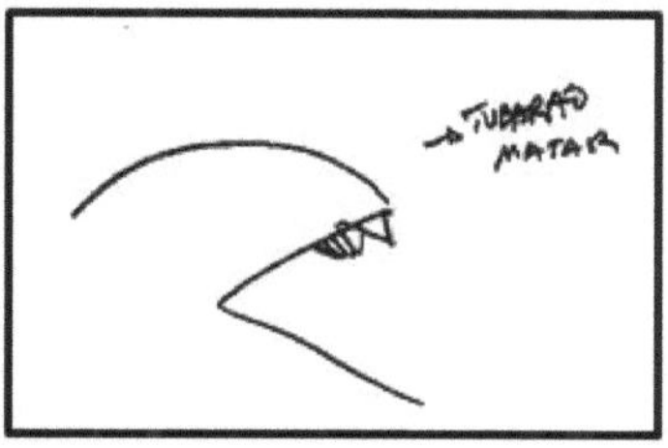

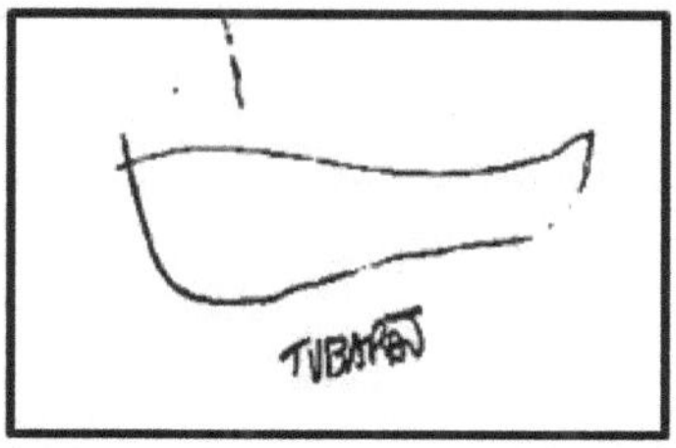

Source: Joca Source: Joca

In his drawings, Joca repeated the figure of a shark. First, only the open mouth, with a few teeth, the absence of various parts of the face and the total absence of the body. Later, he represented the shark only through its body, without joining it to its head. This shows how things are going on inside him, the lack of connections, of integration, whether from a cognitive or emotional point of view.

When he depicts the dinosaur, he manages to go a little further, to create an outline of the animal, including the head, although without the details of the face. The shark, whose jaw is emphasized, with some very sharp teeth, is reminiscent of aggression, in a very primitive way and disconnected from the rest of the body. For a few moments, this is probably how he sees himself. As a result of the researcher's mediation, empathy and the provision of a situation conducive to expression, the drawings evolve, so that the dinosaur takes on a body.

Figure 14 - Dinosaur

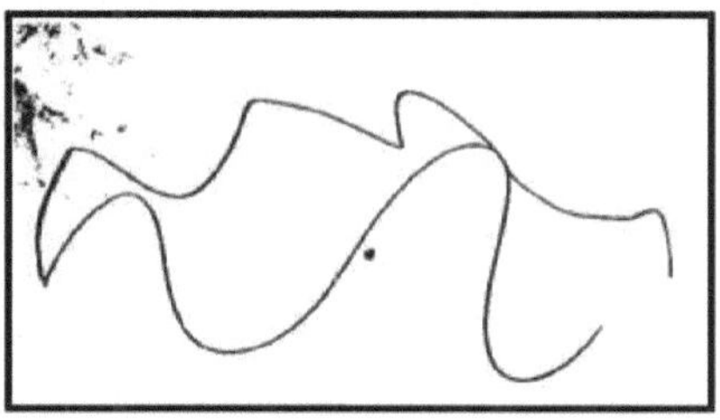

Source: Joca

Figure 15 represents a "human being surrounded by animals, with lots of gorillas and monkeys", according to Joca. In this context, we can see the issue of loneliness, even though there are animals present. As human beings are "surrounded" by animals, it is plausible that there is a fear of contact between people. According to the teacher, there are difficulties in the interaction between Joca and Pedro, especially when one of them is not very well. They are nervous and aggressive towards each other.

Figure 15 - Island with gorillas and monkeys

Source: Joca

As already mentioned, Joca is not historically used to living together. At home, until he left school, he had very limited contact, which could have caused him a lot of anguish. For him, any potential environment represents danger. He sees himself as very small, in a tangle, a bit lost, "without a floor", as shown in the drawing. According to the teacher:

"I think, I have to show you later a recording I made that was presented in a case study of his at the HTPC, we did a study of his case and proposed activities, then I asked the management, then as I said, if we ask they offer to help? Joca is such a child... his family is very complicated. He lives with his grandmother and grandfather, and they both look after a girl, a girl who is in a vegetative state, so as well as the girl there's Joca who has autism. Joca's hand also has a disability, but I can't say which one, and his father is a bit absent, from what Joca tells me, he doesn't talk, he starts to... hitting. The parents and grandparents are neighbors and there are no other children nearby. He plays at school and at home he plays video games. His social problems mean that we have to invest in this, but on the contrary, he's separated from it, he doesn't have it. And the school ends up being in charge of functions that would also be maternal and paternal." P3A

Like Pedro, Joca is reluctant to do activities and when he "explodes" he hits the children. Mainly because of his aggressiveness, he has a tutor who accompanies him during school hours. The boy has difficulties socializing and "picks on" Pedro, making the situation very complicated when the two of them get into disagreements, as they are "unpredictable". Due to the situation he finds himself in, his family is unable to invest in his socialization process, the boy "has no social life, no neighbors and the only activity at home is playing video games." Faced with these extreme difficulties, even though the teacher has a specialization in special education, she asked the school administrators for help, so that she could get a slot in the school meetings to explain the situation and together devise activities for him, with the help of the resource room teacher. The teacher was very tired, frustrated and exhausted.

During the classroom observations, Joca says that he really enjoys playing video games, because there is no one at home for him to play with, but his games are very violent, and he emphasizes this by saying that he likes to kill all his enemies. These aspects can be correlated with the fact that he also really likes Fred Kruger[7] , because it's a horror movie. Another fact that the researcher noticed is that when he suffers some kind of sudden sentimental change that causes him some stress, his repetitive and stereotyped movements increase significantly and as a result he punches his hand quite hard, moves back and forth and looks very nervous.

Unlike Pedro, who internalizes his feelings in a self-defensive move, Joca reverberates his feelings, whether it's good or bad, attacking is the way he's found to communicate. This makes it easier to change these dynamics because every time he attacks Lugar, and he returns it with affection, he changes his perception dynamics. On the other hand, Pedro, who creates an environment of illusions, in which he thinks that the Place itself is attacking him, when in fact they are just thoughts and feelings created by his own psyche, finds it more difficult to change these aspects because they are more deeply rooted.

6. To conclude

In a general analysis of the two cases, we highlight the fact that both had a traumatic relationship at the beginning of their childhood, which makes it difficult for these children to establish a relationship of belonging to the place, reverberating in constant fights and feelings of exclusion. However, aspects of true *self* have been offered to these subjects as a constant attempt by some school professionals, especially the teacher, to find methodologies and ways of dealing with these children, as well as their acceptance by the other pupils, who interact, play and try to help them in some way.

However, no matter how hard the school tries (albeit with its flaws), the boys have created such a strong defense system against frustration in relationships that they don't allow themselves to live well at school, so that they can't grow in the conflicts they have with the other children, even with the mediation of the teachers, which doesn't always seem to be the most appropriate, as Pedro's grandmother said. They believe that they don't deserve

7 Freddy Krueger is a fictional character from the horror film series "A Nightmare on Elm Street." Freddy is a child killer from the fictional Springwood, Ohio, who, after being burned by vengeful parents, starts attacking teenagers in their dreams, killing them in the real world as a matter of course.

love and find it difficult to receive it for fear of weakening the defenses they have created to protect themselves.

Although for these children the relationship with place is still very fragile, when it is built in a congruent and solid way, it can help immensely in this search for their true self and in the possibility of building a relationship with other Places based on subjective flows.

It is important, in this sense, that the Place for these children is conceived as synonymous with care, through health promotion practices that can offer them support, so that they are psychically welcomed in a sufficiently good way. That the child can love and hate, so that they can experience guilt and restitution, that they can relate to people who can survive their retaliation and aggression with the possibility of making amends without the risk of breaking ties and, finally, someone to take responsibility for their care (GUIMARAES; LUZ, 2015).

Bibliographical references

CORRÊA, Roberto Lobato. Socio-spatial differentiation, scale and spatial practices. **cidades**, v. 4, n. 6, 2007, p. 62-72.

GUIMARAES, Leandro Buzzo Mourâo; Luz. Constructing the geographical place of students with autism spectrum disorder in public educational institutions: contributions from psychoanalysis. **Revista Geografia em Atos**. Special Issue. Vol 2. N.2.2015

GUIMARAES, Raul Borges. **Regionalization of health in Brazil: from the scale of the body to the scale of the nation**. Sâo Paulo, University of Sâo Paulo, School of Public Health, 2008. (Free Teaching Thesis)

OLIVEIRA, A.A.S.; LEITE, L.P. Building an inclusive educational system: a political-pedagogical challenge. **Revista Ensaio: Evaluation and Public Policies in Education**. Vol. 15, issue 57, Cesgranrio/RJ, 2007.

SPOSITO, Maria Encarnaçâo. Production of urban space: scales, differences and socio-spatial inequalities. In: **The production of urban space: agents and processes, scales and challenges.** CARLOS, Ana Fani; SOUZA, Marcelo Lopes de; SPOSITO, Maria Encarnaçâo Beltrâo (eds.). Sâo Paulo: Editora Contexto, 2011. p. 53-74

ZOGHBI, A.C.P.; MATTOS, E.H.C. de; ROCHA, F.F.; ARVATE, P.R. Measuring the performance and efficiency of state spending on primary and secondary education. **Estudos Econômicos** v. 39, n. 4, p. 785-809, 2009.

Annexes

1. Geography and Psychoanalysis: a path of possibilities

To the extent that we experience the world differently, we feel and live in different worlds. The Universe is full of subjects going through the same movements in the same environment, but carrying within themselves and projecting around them universes as mutually remote as constellations" (Mc Geanchan, 2014, translation).

In order to think about and get involved with new geographical approaches, we first have to think about the problem from a geographical point of view. In this sense, we ask ourselves: Why study processes relating to psychic life within Geography? This question is pertinent because for a long time this knowledge was considered beyond the limits of geographic knowledge (PARR; DAVIDSON, 2014), as was the study of the body and its corporeality - as highlighted by Alves (2010), and of emotions, based on Bondi (1999). These new perspectives are extremely necessary with the great global advances and new problems faced by geographers and other social sciences, which make us think from other perspectives, from interdisciplinarity, in which we highlight psychic processes and their relationship with Place.

Bondi (1999), a geographer engaged in feminist studies, began to theorize in an article, with rather sentimental effect, what we could call a rapprochement between psychoanalysis and human geography. Through simple but valuable ideas, she managed to connect these two theoretical fields, opening up space for debate about the gaps and possible interlocutions found. In his reports, he states that "*At this stage, all I can offer is a suggestion rather than an exhaustively in-depth argument, but I nevertheless suggest that human geography has forgotten or neglected the links with the humanistic psychotherapy movement (BONDI, 1999, p.229, translation)*"

The studies carried out by Bondi (1999) derive from two theoretical approaches to geography expressed by Freud, the concepts of transference[8] (1912) and ego[9] (1927). Despite using post-Freudian authors

[8] 8 The concept of Transference. Freud (1912) in his book "Psychoanalytic Techniques" theorized that each individual, through the combined action of their innate disposition and the influences suffered during their early years, achieved their own specific method of conducting themselves in life, and when the psychoanalyst returns to the individual's past to try to understand what happened to their life, the patient uses emotions, desires, impressions experienced in childhood and expressed today. It is important to note,

according to Buck and Santos (2009), that this process can take place between student and teacher. Freudians like Lacan, whose studies have shown a still very organicist view of psychoanalysis, describe this scientific field with an image of the decentered subject, with unconscious impulses, ideas and desires, as profoundly at odds with any sense of coherence or agency, as radically incapable of knowing.

Today, authors from various countries such as Kingsbury, P. (2003, 2009); Sibley, D. (2003); Bingley (2003), A.F; Callard, F.; (2003); Bondi, L (2003,2005,2006); Davidson, J. (2003,2005,2007); Gold, J.R. (2009), among many others, have looked at how the theory of the psyche and theories of the *self* can serve to connect the unconscious, emotions and the social and material world through psychological measurement studies in spatial behaviors, emotional geographies, cognitive studies, among other works.

The importance of this strand in geographic studies led the journal Social and Cultural Geography to launch an issue in 2003 (vol.4 ed.3) and, later, in 2015, the book Psychoanalytic Geographies, which reverberates this body of studies accumulated since then. Callard (2003), one of the participants in this issue of the journal, reports in her studies that social and cultural geographies, especially since the turn of the century, have used psychoanalytic authors such as Freud, Lacan, Winnicott, Klein and others to deepen subjectivity and psychic processes in spatial analysis.

Parr and Davidson (2014) emphasize the importance of the urban site in psychic processes, which have shaped and disturbed human beings, creating a complexity in the construction and production of Places. With this effect, psychoanalytic geography, in Coverley's view (2006), focuses on the behavioral impact of urban space, closely associated with the past of the city and the subject, exposing obscure Places (Non-Places) and marginalized Places, i.e. those that are not given importance in scientific studies. *"For the totalizing gaze of the observer sees the city as a homogeneous whole, an anonymous urban space that sees no place for individual or separate identities and that erases or suppresses the personal and the local (COVERLEY, 2006, p. 105) "*

or in an interview in qualitative research.

[9] Freud (1927) considers that the human psyche is made up of the unconscious and the conscious, proposing three concepts to explain this dynamic: Id, Ego and Superego. The Id is linked to the pleasure principle, the most archaic of human satisfactions. The Ego functions by avoiding the pain resulting from social coexistence, it mediates between the Id and the relationships that take place in the external world, it is linked to reality, constantly fighting in an attempt to establish control over the instincts that are in the Id.

The ego also seeks appropriate ways of fulfilling the desires of the Id, so as not to expose the individual to emotional/social conflicts which could have disastrous consequences for their life. The Superego represents morality, the rules of coexistence represented by parents through culture; it represents an ego-ideal.

Bibliographical references

BINGLEY, A.F. In here and out there: sensations between self and landscape. Social and Cultural Geography , vol. 4(3), 2003.p. 329-45.

BONDI, Liz. Stages on Journeys: Some Remarks about Human Geography and Psychotherapeutic Practice, **The Professional Geographer**. vol. 51:1. 1999. p.11-24

. Empathy and identification: conceptual resources for feminist fieldwork.

ACME: **International Journal of Critical Geography ,** vol 2, 2003. p. 64-76.

. The place of emotions in research: from partitioning emotion and reason to the emotional dynamics of research relationships. In: **Emotional Geographies**, J. Davidson, L. Bondi and M. Smith. Aldershot: Ashgate (org).2005. p. 231-46.

COVERLEY, Merlin. **Psycogeography.** Harpenden: Pocket Essentials.2006. p. 158]

DAVIDSON, J.. Phobic Geographies: The Phenomenology and Spatiality of Identity Aldershot: **Ashgate**. 2003

DAVIDSON, J., BONDI, L. and SMITH, M. . Emotional Geographies . Aldershot: **Ashgate.** 2005

DAVIDSON, Joyce; PARR, Hester. Geographies of Psychic Life. In: **Psychoanalytic Geographies,** KINGSBURY, Paul; PILE, Steve (eds). United States: Ashgate Publishing Limited. 2014. p. 143-159 (online).

KINGSBURY, P. Psychoanalytic approaches, in A Companion to Cultural Geography , edited by J.S. Duncan, N.C. Johnson and R. Schein. Blackwell: **Oxford**, 2003. p. 10820.

. Psychoanalysis, in International Encyclopedia of Human Geography , edited by R. Kitchin and N. Thrift. Oxford: **Elsevier**. 2009. p. 480-86.

SIBLEY, D. 2003. Psychogeographies of rural space and practices of exclusion.

Country Visions , edited by P. Cloke. Prentice Hall: **Harlow**, 218-31.

MCGEACHAN, Cheryl . "Worlding" Psychoanalytic Insights: Unpicking R.D. Laing's Geographies. In: **Psychoanalytic Geographies**, KINGSBURY, Paul; PILE, Steve. United States: Ashgate Publishing Limited. 2014. p. 114-126 (online).

2. Approval of the survey Municipal Schools of Presidente Prudente -SP

Município de Presidente Prudente
Secretaria Municipal de Educação
Gabinete da Secretária

Presidente Prudente, 18 de Fevereiro de 2016.

AUTORIZAÇÃO

Diante do parecer favorável da Comissão de Estágios, Projetos de Pesquisa e Extensão, autorizamos o desenvolvimento do trabalho **"Promoção da saúde e inclusão escolar: Produção do lugar social de crianças e jovens com transtorno do Espectro Autista (TEA) nas escolas municipais de Presidente Prudente-SP"**, a ser realizada pelo acadêmico Leandro Buzzo Mourão Guimarães, sob orientação da Profª Dr. Raul Borges Guimarães.

Lembramos que esta Secretaria deverá receber cópia da finalização da pesquisa antes que esta seja proclamada e publicada.

Por ser verdade, firmo o presente.

Ondina [illegible]
Secretária Municipal de Educação

SECRETARIA MUNICIPAL DE EDUCAÇÃO

3. Approval by the Social Sciences Ethics Committee

UNESP - FACULDADE DE
CIÊNCIAS E TECNOLOGIA DO
CAMPUS DE PRESIDENTE

Plataforma Brasil

PARECER CONSUBSTANCIADO DO CEP

DADOS DO PROJETO DE PESQUISA

Título da Pesquisa: Promoção da saúde e inclusão escolar:Produção do lugar social de crianças e jovens com Transtorno doEspectro Autista em escolas municipais de Presidente Prudente - SP

Pesquisador: Raul Borges Guimarães

Área Temática:

Versão: 3

CAAE: 55005416.2.0000.5402

Instituição Proponente: UNIVERSIDADE ESTADUAL PAULISTA JULIO DE MESQUITA FILHO

Patrocinador Principal: FUNDACAO DE AMPARO A PESQUISA DO ESTADO DE SAO PAULO

Conclusões ou Pendências e Lista de Inadequações:

O projeto aborda temática relevante e de impacto. Atende todos os critérios para a sua aprovação.

Considerações Finais a critério do CEP:

Em reunião realizada no dia 14.10.2016, o Comitê de Ética em Pesquisa da Faculdade de Ciências e Tecnologia - Unesp - Presidente Prudente, em concordância com o parecerista, considerou o projeto APROVADO.

Obs: Lembramos que ao finalizar a pesquisa, o (a) pesquisador (a) deverá apresentar o relatório final.

MIX
Papier aus verantwortungsvollen Quellen
Paper from responsible sources
FSC® C105338

Printed by Books on Demand GmbH, Norderstedt / Germany